SpringerBriefs in Computer Science

Series Editors

Stan Zdonik
Peng Ning
Shashi Shekhar
Jonathan Katz
Xindong Wu
Lakhmi C. Jain
David Padua
Xuemin (Sherman) Shen
Borko Furht
V.S. Subrahmanian
Martial Hebert
Katsushi Ikeuchi
Bruno Siciliano

For further volumes:
http://www.springer.com/series/10028

Peter He • Lian Zhao • Sheng Zhou • Zhisheng Niu

Radio Resource Management Using Geometric Water-Filling

 Springer

Peter He
Department of Electrical
and Computer Engineering
Ryerson University
Toronto, ON, Canada

Lian Zhao
Department of Electrical
and Computer Engineering
Ryerson University
Toronto, ON, Canada

Sheng Zhou
Department of Electronic
Engineering
Tsinghua University
Beijing, People's Republic of China

Zhisheng Niu
Department of Electronic
Engineering
Tsinghua University
Beijing, People's Republic of China

ISSN 2191-5768   ISSN 2191-5776 (electronic)
ISBN 978-3-319-04635-8   ISBN 978-3-319-04636-5 (eBook)
DOI 10.1007/978-3-319-04636-5
Springer Cham Heidelberg New York Dordrecht London

Library of Congress Control Number: 2014931367

Printed on acid-free paper

Springer is part of Springer Science+Business Media (www.springer.com)

Preface

Radio resource management (RRM) has been an important task for wireless communication systems. With the emergence of more advanced and more complicated systems, such as cognitive radio, nodes with energy harvest capacities (green communications), and the application of multiple-input multiple-output (MIMO) technology, RRM encounters more difficulties and challenges to optimize system performances. Due to the specific structure of communication systems, water-filling (WF) plays an important role in RRM. This book introduces the fundamental theory and development of WF algorithms. A geometric water-filling (GWF) approach is presented and compared with the conventional WF algorithms. It is shown that GWF provides more insights into solutions and problems. It can break through the limitations of the conventional WF to solve more complicated optimization problems. The applications of the proposed GWF to solve RRM problems in advanced communication systems, e.g., multiple-input multiple-output communication systems, cognitive radio communication systems, and green communication systems, are investigated in this book. Efficient algorithms are presented to achieve optimal resource allocation. Since the authors are active in RRM research, some of their recently published results, as the above-mentioned effective algorithms, are presented in this book. The book addresses important emerging topics in RRM of wireless system design in more detail. No system design is possible without understanding the underlying channels, therefore Chap. 1 focuses on this topic. Many modern radio resource systems use water-filling as the underlying tool. Chapter 2 discusses this important technology in great detail. RRMs of both single-user and multi-user cases, and optimality of the RRM issues in system design, are presented in Chap. 3 for the MIMO system. Chapter 4 considers the scheduling problem under the cognitive radio (CR) network, i.e., allocation of resources to multiple users in a wireless system under CR network, an issue of critical importance in a multi-user system. Chapter 5 considers the optimal design of RRM in harvested energy in order to achieve maximum throughput, for green communications. Every chapter offers examples in a separate section, which unfolds the potential power of developing water-filling.

Toronto, Ontario, Canada   Peter He, Lian Zhao
Bejing, China   Sheng Zhou, Zhisheng Niu
August 2013

Acknowledgements

The authors sincerely acknowledge the support from Natural Sciences and Engineering Research Council (NSERC) of Canada under Grant numbers RGPIN/293237-2009, National Basic Research Program of China (973 GREEN: 2012CB316001) and National Science Foundation of China (No.61201191, No.60925002).

A very special thanks to Prof. Sherman Shen who made this book possible.

Contents

Acronyms

$A(\cdot)$	Mapping
$\mathbb{C}(H,P)$	Channel capacity
$\mathbb{C}^n$	Unitary space with dimension n
$\mathbb{C}^{m\times n}$	Set of $m \times n$ complex matrices
$\det(B)$ or $|B|$	Determinant of the square matrix B
$E(\xi)$	Expected value of the random variable ξ
f_ξ	Probability density function of random variable ξ
H	Channel matrix
$H^\dagger$	Conjugate transpose of matrix H
$H_i^\dagger$	Conjugate transpose of matrix H_i
H_i^T	Transpose of matrix H_i
$\mathbb{H}(\xi)$	Differential entropy
$\mathbb{H}(\xi|\eta)$	Conditional entropy
I_r	Identity matrix with dimension r
$\mathbb{I}(\xi;\eta)$	Mutual information
log	Natural logarithm
$M \succeq 0$	Matrix M being positive semi-definite
$\mathrm{Tr}(M)$	Trace of matrix M
$\lfloor \ \rfloor$	Floor function
$\lceil \ \rceil$	Ceiling function
KKT	Karush-Kuhn-Tucker
WFA	Water-filling algorithm
IWFA	Iterative water-filling algorithm
MIMO	Multiple input multiple output
MAC	Multiple access channel
BC	Broadcast channel
CR	Cognitive radio
EH	Energy harvest
GWF	Geometric water-filling

Chapter 1
Introduction

The world is demanding more from wireless communication services now than ever before. Many advanced wireless communication techniques, such as multiple-input multiple-output (MIMO), cognitive radio (CR), energy harvesting (EH) communications have attracted lots of research attention. For a wireless network operating in a fading environment, power and bandwidth are precious radio resources which need careful planning. Reducing power consumption to satisfy the target QoS requirement leads to enlarged system capacity and prolonged battery life. With the evolution to the more complicated wireless communication systems, the issues of optimal radio resource management problems have become more and more important to achieve overall optimal system design.

Water-filling has been an important algorithm for radio resource management (RRM) problem in wireless communication systems. In this book, we present a new geometric water-filling (GWF) algorithm, and explore its application in solving RRM problems for the advanced wireless communication systems. In the remaining of this chapter, we will first review the advances in MIMO, CR, EH communication systems, and water-filling algorithm. In the subsequent chapters, we will further discuss the application of water-filling algorithm to solve RRM problems in these advanced communication systems.

1.1 Multiple Input Multiple Output System

The use of multiple antennas at the transmitter and receiver, i.e., Multiple-Input Multiple-Output (MIMO) technology, constitutes a breakthrough ([1], page 1) in the design of wireless communication systems. MIMO technology is now at the core of several existing and emerging wireless standards ([1], page 18). Exploiting multipath scattering, MIMO techniques have delivered significant performance enhancements in terms of data transmission rate and interference reduction on

P. He et al., *Radio Resource Management Using Geometric Water-Filling*, SpringerBriefs in Computer Science, DOI 10.1007/978-3-319-04636-5_1, © The Author(s) 2014

point-to-point links. In Chaps. 3–4 of this book, we focus our attention on multi-user MIMO systems. In particular, we consider the design of multi-user systems so as to enable operation at rates approaching the fundamental limits.

1.2 Cognitive Radio Network

To improve the spectrum utilization, cognitive radio (CR) has been proposed that can provide adaptability for wireless transmission on licensed spectrum. A cognitive radio (CR) network refers to a secondary network operating in a frequency band originally licensed/allocated to a primary network consisting of one or multiple primary users (PUs). A fundamental challenge for realizing such a system is to ensure the quality of service (QoS) of the PUs as well as to maximize the throughput of the secondary users (SU). The proposed research will include system requirement of CR networks.

1.3 Energy Harvest Technology

In recent years, energy harvest in green communications has attracted a lot of research attention due to its environment friendly features. One possible technique to overcome the limitation of battery lifetime is to harvest energy from the environment. In such systems, harvesting energy has become a preferred choice for supporting green communications. However, harvesting energy depends on natural conditions and thus is random over time. It owns causality of power usage, corresponding to the dependence of harvested energy on time development. As a result, the energy from this constraint is often considered to regulate the overall energy flow of the system. The optimal radio resource allocation problem to maximize system throughput turns out to be more complicated.

1.4 Water-Filling Algorithm

In many engineering problems, water-filling plays an important role in radio resource management. For communications, it stems from a class of the problems of maximizing the mutual information between the input and the output of a channel with parallel independent sub-channels. With water-filling, more power is allocated to the channels with higher gains to maximize the sum of data rates or the capacity of all the channels. The solution to this class of the problems can be interpreted by a vivid description as pouring limited volume of water into a tank, the bottom of which has the stair levels determined by the inverse of the sub-channel gains.

The conventional way to solve the water-filling problem is to solve the Karush-Kuhn-Tucker (KKT) conditions [2], and then find the water-level(s) and the solutions. In our paper [3], we proposed a water-filling algorithm from geometric approach (GWF) [3]. GWF provides closed form solutions to the power allocation problem and avoids the complexity to solve the KKT conditions with non-linear equations. Due to complexity of solving the KKT conditions of the problem with multiple variables, the GWF is easier to compute than the conventional water-filling and reveals more useful information. GWF has also been extended to solve more general and more complicated power allocation problems.

A chapter of Appendix is included at the end of this book to assist the understanding of the contents in this book.

References

1. E. Biglieri, R. Calderbank, A. Constantinides, A. Goldsmith, A. Paulraj and H. V. Poor, MIMO Wireless Communications, Cambridge University Press, Cambridge, 2007.
2. D. P. Bertsekas, Nonlinear Programming, 2nd Edition, Athena Scientific, Nashua, 1999.
3. P. He, L. Zhao, S. Zhou and Z Niu, "Water-filling: A geometric approach and its application to solve generalized radio resource allocation problems," IEEE Transactions on Wireless Communications, vol. 12, pp. 3637–3647, 2013.

Chapter 2
Geometric Water-Filling in RRM

In this chapter, we introduce water-filling algorithms to solve power allocation problems. Two water-filling approaches are presented. One is the conventional water-filling (CWF); and the other one is the proposed geometric water-filling (GWF). GWF is further extended to efficiently solve a class of power allocation problems with more complex structure which owns upper bounds of the power variables. Computational complexities are investigated.

2.1 Problem Statement and Water-Filling

The water-filling problem can be abstracted and generalized into the following problem: given $P > 0$, as the total power or volume of the water; the allocated power and the propagation path gain for the ith channel are given as s_i and a_i respectively, $i = 1, \ldots, K$; and K is the total number of channels. Letting $\{a_i\}_{i=1}^{K}$ be a sorted sequence, which is positive and monotonically decreasing, find that

$$\max_{\{s_i\}_{i=1}^{K}} \quad \sum_{i=1}^{K} \log(1 + a_i s_i)$$
$$\text{subject to} : 0 \leq s_i, \forall i; \qquad (2.1)$$
$$\sum_{i=1}^{K} s_i = P.$$

Since the constraints are that (i) the allocated power to be nonnegative; (ii) the sum of the power equals P, the problem (2.1) is called the water-filling (problem) with sum power constraint.

To find the solution to problem (2.1), we usually start from the Karush-Kuhn-Tucker (KKT) conditions of the problem, as a group of the optimality conditions, and derive the system (2.2) below from the KKT conditions,

P. He et al., *Radio Resource Management Using Geometric Water-Filling*, SpringerBriefs in Computer Science, DOI 10.1007/978-3-319-04636-5_2, © The Author(s) 2014

$$\begin{cases} s_i = \left(\mu - \frac{1}{a_i}\right)^+, \text{ for } i = 1, \dots, K, \\ \sum_{i=1}^{K} s_i = P, \\ \mu \geq 0, \end{cases} \qquad (2.2)$$

where $(x)^+ = \max\{0,x\}$. μ is the water level chosen to satisfy the power sum constraints with equality ($\sum_{i=1}^{K} s_i = P$). The solution to (2.2) is referred as a solution of the CWF problem (2.1).

It can be seen that the implied system (2.2) has been used to find the optimal solution. The existence of its Lagrange multipliers and the implication mentioned above determine that enumeration can be utilized to find the water level μ. In [1], how to solve the problems has been discussed extensively. Complexity of the non-geometric approach to solve the problem (2.1) will be discussed in Sect. 2.5. In the sequel of the chapter, when water-filling problem is mentioned, the power sum constraint is always included.

2.2 Proposed Geometric Water-Filling Approach

In this chapter, we propose a novel approach to solve problem (2.1) based on geometric view. The proposed Geometric Water-Filling (GWF) approach eliminates the procedure to solve the non-linear system for the water level, and provides explicit solutions and helpful insights to the problem and the solution.

Figure 2.1a–c give an illustration of the proposed GWF algorithm. Suppose there are 4 steps/stairs ($K = 4$) with unit width inside a water tank. For the conventional approach, the dashed horizontal line, which is the water level μ, needs to be determined first and then the power allocated for each stair (water volume above the stair) is solved.

Let us use d_i to denote the "step depth" of the ith stair which is the height of the ith step to the bottom of the tank, and is given as

$$d_i = \frac{1}{a_i}, \text{ for } i = 1, 2, \dots, K. \qquad (2.3)$$

Since the sequence a_i is sorted as monotonically decreasing, the step depth of the stairs indexed as $\{1, \cdots, K\}$ is monotonically increasing. We further define $\delta_{i,j}$ as the "step depth difference" of the ith and the jth stairs, expressed as

$$\delta_{i,j} = d_i - d_j = \frac{1}{a_i} - \frac{1}{a_j}, \text{ as } i \geq j \text{ and } 1 \leq i, j \leq K. \qquad (2.4)$$

Instead of trying to determine the water level μ, which is a real nonnegative number, we aim to determine water level step, which is an integer number from 1 to K, denoted by k^*, as the highest step under water. Based on the result of k^*, we can write out the solutions for power allocation instantly.

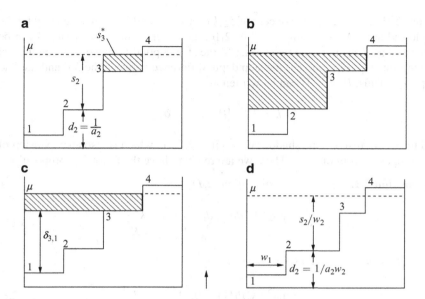

Fig. 2.1 Illustration for the proposed geometric water-filling (GWF) algorithm. (**a**) Illustration of water level step $k^* = 3$, allocated power for the third step s_3^*, and step/stair depth $d_i = 1/a_i$. (**b**) Illustration of $P_2(k)$ (*shadowed area*, representing the total water/power above step k) when $k = 2$. (**c**) Illustration of $P_2(k)$ when $k = 3$. (**d**) Illustration of the weighted case

Figure 2.1a illustrates the concept of k^*. Since the third level is the highest level under water, we have $k^* = 3$. The shaded area denotes the allocated power for the third step by s_3^*.

In the following, we explain how to find the water level step k^* without the knowledge of the water level μ. Let $P_2(k)$ denote the water volume above step k or zero, whichever is greater. The value of $P_2(k)$ can be solved by subtracting the volume of the water under step k from the total power P, as

$$P_2(k) = \left\{ P - \left[\sum_{i=1}^{k-1} \left(\frac{1}{a_k} - \frac{1}{a_i} \right) \right] \right\}^+ = \left\{ P - \left[\sum_{i=1}^{k-1} \delta_{k,i} \right] \right\}^+, \text{ for } k = 1,\ldots,K.$$

(2.5)

Due to the definition of $P_2(k)$ being the power (water volume) above step k, it cannot be a negative number. Therefore we use $\{\cdot\}^+$ in (2.5) to assign 0 to $P_2(k)$ if the result inside the bracket is negative. The corresponding geometric meaning is that the kth level is above water. Note a reminder of the definition of a special case for the summation is:

$$\sum_{i=m}^{n} b_i = 0, \text{ as } m > n.$$

(2.6)

Figure 2.1b, c illustrate the concept of $P_2(k)$ for $k = 2$ and $k = 3$ respectively by the shadowed area. As an example of Fig. 2.1c, the water volume under step 3 can be expressed as the sum of the two terms: (i) the step depth difference between the 3rd and the 1st step, $\delta_{3,1}$, and (ii) the step depth difference between the 3rd and the 2nd step, $\delta_{3,2}$. Thus, $P_2(k = 3)$ can be written as

$$P_2(k = 3) = [P - \delta_{3,1} - \delta_{3,2}]^+$$

and the above result is the shadowed area in Fig. 2.1c, which is also an expansion of the composite form of (2.5). Then, we are ready to have the following proposition:

Proposition 2.1. *The explicit solution to (2.1) is:*

$$s_i = \begin{cases} s_{k^*} + (d_{k^*} - d_i) & 1 \leq i \leq k^* \\ 0, & k^* < i \leq K, \end{cases} \tag{2.7}$$

where the water level step k^* is given as

$$k^* = \max \left\{ k \middle| P_2(k) > 0, \ 1 \leq k \leq K \right\} \tag{2.8}$$

and the power level for this step is

$$s_{k^*} = \frac{1}{k^*} P_2(k^*). \tag{2.9}$$

It is easy to interpret Proposition 2.1 from Fig. 2.1. The first step of the proposed approach is to find the water level step k^*. From Fig. 2.1, we can find that $k = 3$ is the maximal index that makes $P_2(k)$ greater than zero. Therefore, based on (2.8), $k^* = 3$ can be determined. Then the power at this step s_{k^*} can be determined based on (2.9). For those steps with index higher than k^*, no power is assigned. For those steps with index lower than k^*, their power levels are obtained by adding s_{k^*} with the corresponding level depth difference with the k^*th step as shown in (2.7).

Proposition 2.1 provides an explicit constructed solution rather than the implicit solution. The procedure eliminates solving the nonlinear equation as shown in (2.2) and the real number water level μ. The proof of the optimality of the solution will be left to the next subsection when we discuss the weighted case.

2.3 Generalize to Weighted Case

For the weighted case, the generalized problem can be stated as: given $P > 0$, as the total power or volume of the water; the allocated power and the propagation path gain for the ith antenna are given as s_i and a_i respectively, $i = 1, \ldots, K$; and K is the total number of the transmit antennas. Furthermore, the weighted coefficients $w_i > 0, i \in \{1,\ldots,K\}$, and $\{a_i w_i\}_{i=1}^{K}$ being monotonically decreasing, find that

$$\max_{\{s_i\}_{i=1}^{K}} \quad \sum_{i=1}^{K} w_i \log(1 + a_i s_i)$$
$$\text{subject to} : 0 \leq s_i, \forall i;$$
$$\sum_{i=1}^{K} s_i = P. \tag{2.10}$$

Using the proposed geometric approach, we can extend the geometric relation for the weighted case as shown in Fig. 2.1d to obtain the corresponding solution to (2.10).

In Fig. 2.1d, the width of the ith stair/step is denoted as w_i. The value of $1/a_i$ denotes the volume under the ith step to the bottom of the tank. Hence, the step depth of the ith step is given as

$$d_i = \frac{1}{a_i w_i}, \quad i = 1, \cdots, K. \tag{2.11}$$

Then, $P_2(k)$, the water volume above step k, can be obtained using the similar approach as in the previous subsection considering the step depth difference and the width of the stairs as,

$$P_2(k) = \left[P - \sum_{i=1}^{k-1} (d_k - d_i) w_i \right]^{+}, \text{ for } k = 1,\ldots,K. \tag{2.12}$$

As an example in Fig. 2.1d, the water volume above step 1 and below step 3 with the width w_1 can be found as: the step depth difference, $(d_3 - d_1)$ multiplying the width of the step, w_1. Therefore, the corresponding $P_2(k = 3)$ can be expressed as,

$$P_2(k = 3) = [P - (d_3 - d_1)w_1 - (d_3 - d_2)w_2]^{+},$$

which is an expansion of (2.12). Then we have the following proposition.

Proposition 2.2. *The explicit solution to (2.10) is:*

$$\begin{cases} s_i = [\frac{s_{k^*}}{w_{k^*}} + (d_{k^*} - d_i)]w_i, \text{ as } 1 \leq i \leq k^*; \\ s_i = 0, \text{ as } k^* < i \leq K, \end{cases} \tag{2.13}$$

where

$$k^* = \max \left\{ k \middle| P_2(k) > 0, \ 1 \leq k \leq K \right\} \tag{2.14}$$

and the power level for this step is

$$s_{k^*} = \frac{w_{k^*}}{\sum_{i=1}^{k^*} w_i} P_2(k^*). \tag{2.15}$$

Proof of Proposition 2.2. System (2.13) implies that

$$\frac{w_{k^*}}{\frac{1}{a_{k^*}} + s_{k^*}} = \frac{w_i}{\frac{1}{a_i} + s_i}, \text{ as } 1 \leq i \leq k^*. \tag{2.16}$$

Let

$$\lambda = \frac{w_{k^*}}{\frac{1}{a_{k^*}} + s_{k^*}}. \tag{2.17}$$

From a geometric view, λ is the reciprocal of water level μ. According to the definitions of k^* and s_{k^*}, for $k^* < i \leq K$, $\frac{w_{k^*}}{\frac{1}{a_{k^*}} + s_{k^*}} > \frac{w_i}{\frac{1}{a_i} + s_i}$ and $s_i = 0$.

Let

$$\sigma_i = \frac{w_{k^*}}{\frac{1}{a_{k^*}} + s_{k^*}} - \frac{w_i}{\frac{1}{a_i} + s_i}. \tag{2.18}$$

Then

$$\begin{cases} \sigma_i > 0, \text{ as } k^* < i \leq K \\ \sigma_i = 0, \text{ as } 1 \leq i \leq k^*. \end{cases} \tag{2.19}$$

Therefore, the following system holds:

$$\begin{cases} \frac{w_i}{\frac{1}{a_i} + s_i} - \lambda + \sigma_i = 0, \text{ as } 1 \leq i \leq K \\ s_i \geq 0, \qquad\qquad \forall i \\ \sigma_i s_i = 0, \qquad\qquad \forall i \\ \sigma_i \geq 0, \qquad\qquad \forall i \\ \sum_{i=1}^{K} s_i = P, \qquad\quad \lambda \in \mathbb{R}. \end{cases} \tag{2.20}$$

By observation, the equation and inequality set above is just a set of the KKT conditions of the problem in Proposition 2.2 and the water level μ is equal to the reciprocal of the Lagrange multiplier λ mentioned above. Note that the Lagrange function of the problem in Proposition 2.2 is

$$L(\{s_i\}, \lambda, \{\sigma_i\}) = \sum_{i=1}^{K} w_i \log(1 + a_i s_i) - \lambda\left(\sum_{i=1}^{K} s_i - P\right) + \sum_{i=1}^{K} \sigma_i s_i. \tag{2.21}$$

Since it is a differentiable convex optimization problem with linear constraints, not only are the KKT conditions mentioned above sufficient, but they are also necessary for optimality. Note that the constraint qualification of the problem (2.10) holds. Proposition 2.2 hence is proved.

Similar to the unweighted case, the first step is to calculate $P_2(k)$, then find the water level step, k^*, from (2.14), which is the maximal index making $P_2(k)$ nonnegative. The corresponding power level for this step, s_{k^*}, can be obtained by applying (2.15). Then for those steps with index higher than k^*, the power level is assigned with zero. For those steps below k^*, the power level is assigned as in (2.13). The first term (s_{k^*}/w_{k^*}) inside the square bracket denotes the depth of the k^*th step to the surface of the water. The second term inside the square bracket denotes the step depth difference of the k^*th step and the ith step. Therefore, the sum inside the square bracket means the depth of the ith step to the surface of the water. When this quantity is multiplied with the width of this step, the volume of the water above this step (allocated power) can be then readily obtained.

With the proposed GWF approach, the weighted problem could be solved straightforwardly, avoiding complicated derivation and calculation. When the weighting factors are set to ones, the corresponding unweighted case is obtained. In the following description of algorithm implementation and proof, we only provide weighted case.

From Proposition 2.2, when k^* is obtained, $P_2(k^*)$ is given. Then it is memorized and only multiplied by a constant to compute s_{k^*}. Thus, how to search k^* is a key point for the proposed GWF and the procedure is stated as follows:

1. Initialize $W_s = 0; P_M = P^* = P; i = 1$.
2. Compute $W_s <= W_s + w_i; P^* <= P^* - (d_{i+1} - d_i)W_s$. Then $i <= i + 1$, where the symbol "$<=$" represents the assignment operation.
3. If $P^* > 0$ and $i \leq K$, $P_M = P^*$, and repeat the step 2; else, output $k^* = i - 1, W_s = W_s - w_i$ and $s_{k^*} = \frac{w_{k^*}}{W_s}P_M$.

We can observe that $\frac{s_{k^*}}{w_{k^*}} + d_{k^*}$ is the water level due to $\frac{s_{k^*}}{w_{k^*}} + d_{k^*} = \frac{s_i}{w_i} + d_i$, for $1 \leq i \leq k^*$.

As an alternative to the enumeration search in the Algorithm GWF, a Fibonacci-like search is possibly used to speed up finding k^* due to (non-increasing) monotonicity of the sequence $\{P_2(k)\}$. Without loss of generality, let Fibonacci approximation ratios be $\frac{1}{3}$ and $\frac{2}{3}$ for searching k^*. The method can be described as:

1st Step.   Assume that $a = 1$ and $b = K$.
2nd Step.   If $a = b$, then $k^* = a$ and go to Step 3 of GWF.
   Else, $a_1 = \lfloor a + \frac{1}{3}(b - a) \rfloor$, $b_1 = \lceil a + \frac{2}{3}(b - a) \rceil$.
3rd Step.   If $P_2(a_1) \leq 0$, then $b = a_1 - 1$ and go to the 2nd Step;
   If $P_2(b_1) > 0$, then $a = b_1$ and go to the 2nd Step;
   If $P_2(a_1) > 0$ and $P_2(b_1) \leq 0$, then $a = a_1, b = b_1 - 1$ and go to the 2nd Step.

The number of loops to search k^* is reduced into a complexity level of $\log_3(K)$.

2.4   Weighted Water-Filling with Individual Peak Power Constraints

In this section, we extend the CWF problem to include individual peak power constraints (WFPP).

The weighted WFPP problem is stated as follows. Given $P > 0$, as the total power or volume of the water; the allocated power and the propagation path gain for the ith antenna are given as s_i and a_i respectively, $i = 1, \ldots, K$; and K is the total number of the transmit antennas. Also, the weights $w_i > 0, \forall i$, and without loss of generality, $\{a_i \cdot w_i\}_{i=1}^{K}$ being positive and monotonically decreasing, find that

$$\max_{\{s_i\}_{i=1}^{K}} \quad \sum_{i=1}^{K} w_i \log(1 + a_i s_i)$$
$$\text{subject to} : 0 \leq s_i \leq P_i, \; \forall i; \quad\quad\quad\quad (2.22)$$
$$\sum_{i=1}^{K} s_i \leq P.$$

Comparing the problem (2.22) with (2.10), the constraint of $0 \leq s_i$ is extended to $0 \leq s_i \leq P_i$, i.e., additional individual peak power constraints, and $\sum_{i=1}^{K} s_i = P$ to $\sum_{i=1}^{K} s_i \leq P$. The problem (2.22) is thus referred to as (weighted) water-filling with sum and individual peak power constraints (WFPP). In this section, we discuss the solution to the WFPP problem.

Proposition 2.2 in Sect. 2.3 provides an explicit solution using geometric view approach. Interestingly, the proposed GWF can be applied to the WFPP problem with some modifications. The following presents an algorithm which is a modification of the above discussed GWF and it is termed as the GWFPP.

For convenience, the expression (2.12) can further be extended into the expression:

$$P_2(i_k) = \left[P - \sum_{t=1}^{|E|-1} \left(d_{i_k} - d_{i_t} \right) w_{i_t} \right]^{+}, \text{for } k = 1, \ldots, |E|, \quad\quad (2.23)$$

where E is a subsequence of the sequence $\{1, 2, \ldots, K\}$, $|E|$ is the cardinality of the set E, so E can be expressed as $\{i_1, i_2, \ldots, i_{|E|}\}$. Especially, if E is taken as the sequence $\{1, 2, \ldots, K\}$, then the extended expression is regressed into the original expression (2.12). Similarly, some corresponding changes in (2.13)–(2.15) are also done (i.e., the subscripts of sequence are replaced with those of the subsequence). For avoiding tediousness, these extended expressions are still labelled as (2.13)–(2.15) in the following statement of Algorithm GWFPP.

Algorithm GWFPP:

Input: vector $\{d_i\}, \{w_i\}, \{P_i\}$ for $i = 1, 2, \ldots, K$, the set $E = \{1, 2, \ldots, K\}$, and P.

1. Utilize (2.13)–(2.15) to compute $\{s_i\}$.
2. The set Λ is defined by the set $\{i | s_i > P_i, i \in E\}$. If Λ is the empty set, output $\{s_i\}_{i=1}^{K}$; else, $s_i = P_i$, as $i \in \Lambda$.
3. Update E with $E \setminus \Lambda$ and P with $P - \sum_{t \in \Lambda} P_t$. Then return to (1) of the GWFPP.

Remark 2.1. Algorithm GWFPP is a dynamic power distribution process. The state of this process is the difference between the individual peak power sequence and the current power distribution sequence obtained by the Algorithm GWF. The control of this process is to use (2.13)–(2.15) of the Algorithm GWF based on the state mentioned above. Thus, a new state for next time stage appears. Therefore, an optimal dynamic power distribution process, the GWFPP, with the state feedback is formed. Since the finite set E is getting smaller and smaller until the set Λ is empty, Algorithm GWFPP carries out K loops to compute the optimal solution, at most.

Similar to the proof of the Proposition 2.2, we can obtain the following conclusion:

Proposition 2.3. *Algorithm GWFPP can provide the optimal solution to the problem (2.22).*

Proof of Proposition 2.3. If the final set E in Algorithm GWFPP is empty, it implies that $\sum_{i=1}^{K} P_i \leq P$. Then it is easy to see the optimal solution $s_i = P_i$, for any i.

If it is non-empty, observing the stricture of (2.22), Proposition 2.3 is easily proved, similarly to the previous one.

2.5 Complexity Analysis

As stated in [1] (Sect. 2.3), the conventional WF algorithm had an exponential worst-case complexity of 2^K, where K is the number of the channels, even though the channel gains had been sorted in decreasing order. Pointing to this case, [1] proposed an improved algorithm with worst-case complexity of K iterations. Since each iteration consists of multiple arithmetic and logical operations, here we use total number of operations as a measure of the complexity level (See [2], Chap. 8).

The CWF approach has a worst-case complexity of K iterations, i.e., total $O(K^2)$ fundamental arithmetic and logical operations under the $2(K+1)$ memory requirement and the sorted parameters $\{w_k a_k\}_{k=1}^{K}$ (e.g. see [3], page 137, for more details).

The proposed GWF algorithm occupies less computational resource. It is seen that it needs K loops at most to search k^* and it needs four arithmetic operations and two logical operations to complete each loop. Thus, the worst-case computational complexity of the proposed solution is $8K+3$ (from the operations of $6K+3+2K$) fundamental arithmetical and logical operations under the $2(K+1)$ memory units to store $\{d_i\}$, $\{w_i\}$, W_s, and P_M.

For the GWFPP, it needs K loops to compute the optimal solution, at most. The required number of operations is, at worst, $\sum_{i=1}^{K}(8i+3) = 4K^2 + 7K$ fundamental arithmetical and logical operations.

Note that the content of this chapter comes partially from [4] and references therein.

References

1. D. Palomar, "Practical algorithms for a family of waterfilling solutions," IEEE Transactions on Signal Processing, vol. 53, pp. 686–695, 2005.
2. C. H. Papadimitriou and K. Steiglitz, Combinatorial Optimization: Algorithms and Complexity, Unabridged edition, Dover Publications, Mineola, 1998.
3. D. Palomar, PhD Thesis: A Unified Framework for Communications through MIMO Channels, Universitat Politecnica De Catalunya, Spain, 2003.
4. P. He, L. Zhao, S. Zhou and Z Niu, "Water-filling: A geometric approach and its application to solve generalized radio resource allocation problems," IEEE Transactions on Wireless Communications, vol. 12, pp. 3637–3647, 2013.

Chapter 3
RRM in MIMO System

In this chapter, we first discuss the single user MIMO system and its optimal power allocation solution. Then, we present two classes of MIMO system models: MAC (multiple access channel) and BC (broadcast channel). Their optimal power allocation computation is provided. More solid and more efficient computation are set up in this book. Finally, optimality of the optimal power allocation policies for the multi-user MIMO MAC and BC is provided. Part of the contents of this chapter are published at [1, 2].

3.1 Model of the Single User MIMO Channel

A single user Gaussian channel with multiple transmitting and/or receiving antennas is considered as follows. We denote the number of transmitting antennas by t and the number of receiving antennas by r. We restrict our discussion to a linear model in which the received vector $y \in \mathbb{C}^r$ depends on the transmitted vector $x \in \mathbb{C}^t$ via

$$y = Hx + z, \tag{3.1}$$

where H is a $r \times t$ complex channel gain matrix and $z \in \mathbb{C}^r$. We assume that vector z is a zero mean circular complex Gaussian noise vector; cf. [3] and Sect. A.1. For any complex matrix or vector, its superscript † denotes the conjugate transpose of the matrix or the conjugate transpose of the vector. Without loss of generality, we assume $E\left[zz^{\dagger}\right] = I_r$, where I_r is an identity matrix with order r and $E\left[\cdot\right]$ denotes the expectation operation. That is, the noise observed at different receivers is independent with each other. The average power of the transmitter is bounded by P, i.e.,

$$E\left[x^{\dagger}x\right] \leq P,$$

or equivalently,

$$\text{Tr}\left(E\left[xx^{\dagger}\right]\right) \le P. \tag{3.2}$$

This second form of the power constraint will be shown to be more useful than the first form in the upcoming discussions.

In a generic wireless communications set up, there are three scenarios [4] for the matrix H:

1. H is deterministic,
2. H is a random matrix, which is chosen according to a probability distribution, and
3. H is a random matrix, but is fixed once it is chosen.

The focus of this thesis is on the first of these cases. This case is normally referred to as the static channel case with full channel state information at transmitter and receiver sides, e.g., [5, 6]. In this chapter, we will use the GWF, which has been mentioned in Chap. 2, with Fibonacci search for this system. This algorithm will also be utilized in the subsequent chapters. Since GWF owns efficient computation, water-filling appearing in this chapter and following chapters means the GWF without confusion. The details will be referred to in the corresponding chapters.

3.1.1   Channel Capacity of the Single User MIMO Channel

In this subsection, we will discuss the channel capacity of the single user MIMO system in (3.1), with perfect channel state information at the transmitter (see, e.g., [4]).

Given the model in (3.1), the channel capacity C is defined as $C \triangleq \max_{p_x} \mathscr{I}(x;y)$, where $\mathscr{I}(x;y)$ is the mutual information between x and y, and p_x is the probability density function of x. Let $S \triangleq E\left[xx^{\dagger}\right]$. As shown in (3.2), the input power constraint can be written as that $\text{Tr}(S) \le P$. The corresponding channel capacity $C(H,P)$ is expressed as:

$$C(H,P) = \max_{p_x}\left\{\mathscr{I}(x;y) \,|\, S \succeq 0, \text{Tr}(S) \le P\right\}. \tag{3.3}$$

Using the argument in Sect. A.2, it can be shown that for the model in (3.1) in which the channel H is deterministic, and the additive noise is Gaussian and, without loss of generality, has a unit variance, the optimal input distribution for the input x is zero-mean and Gaussian and hence the mutual information can be written as $\log\left(\det\left(I_r + HSH^{\dagger}\right)\right)$. Since a zero-mean Gaussian distribution is completely specified by its covariance, the expression in (3.3) can be simplified to

$$C(H,P) = \max_{S}\left\{\log\left(\det\left(I_r + HSH^{\dagger}\right)\right) \,|\, S \succeq 0, \text{Tr}(S) \le P\right\}. \tag{3.4}$$

The expression on the right hand side of (3.4) is a nonlinear semidefinite optimization problem in the complex matrix variable. A direct and effective algorithm is presented in the following section.

3.1.2 Simplification of the Problem of the Optimal Power Allocation

To efficiently solve the problem in (3.4) of the optimal input covariance for the single user MIMO channel, we first simplify the problem; e.g., [3]. Using the singular value decomposition (SVD) of the H, $H = U\Sigma V^{\dagger}$ and the properties of the determinant, we have that

$$\log \det \left(I_r + HSH^{\dagger} \right) = \log \det \left(I_r + \Sigma \hat{S} \Sigma^{\dagger} \right),$$

where $\hat{S} = V^{\dagger}SV$. Using Hadamard's Determinant inequality, it can be shown that we can restrict attention to diagonal $\hat{S}$, say Γ, and any optimal input covariance takes the form $S = V\Gamma V^{\dagger}$, where $\Gamma = \text{Diag}\,(\gamma_i^*)$ and

$$\{\gamma_i^*\}_{i=1}^t = \arg \max_{\{\gamma_i\}_{i=1}^t} \left\{ \sum_{i=1}^t \log\left(1 + \lambda_i \gamma_i\right) \Big| \gamma_i \geq 0, \forall i; \sum_{i=1}^t \gamma_i \leq P \right\}, \qquad (3.5)$$

where γ_i is the i-th element of $\Sigma^{\dagger}\Sigma$. The remaining challenge is to obtain an efficient algorithm for solving (3.5). The problem in (3.5) can be solved using the conventional water-filling (i.e., CWF) procedure. However, the CWF procedure involves enumeration over the number of the diagonal elements of Γ that are to be non zero. The following algorithm is actually GWF with the Fibonacci search to determine this number of active subchannels, and hence is significantly more computationally efficient than the enumerative algorithm.

Since the GWF procedure is repeated many times in iterative water-filling algorithms (IWFA), the reduction in complexity can have a significant impact in practice. As far as we are aware, the following GWF algorithm is the first time in which Fibonacci search has been used in GWF.

The GWF with Fibonacci search is stated as follows:

Algorithm: GWF Algorithm with Fibonacci Search

Step 1: Pre-Processing.   Compute the unitary matrix $U \in \mathbb{C}^{t \times t}$ by the SVD:

$$\Lambda = U^{\dagger}H^{\dagger}HU = \text{diag}\,(\lambda_1, \cdots, \lambda_t).$$

Let $\{\lambda_i\}_{i=1}^t$ be ordered in the monotonically decreasing order; Let

$$\hat{i} \overset{\triangle}{=} \max\{i|\lambda_i > 0\} \leq \min\{t, r\}.$$

Step 2: GWF with Fibonacci Search for (3.5).   For $k = 1, 2, \ldots, \hat{i}$, let

$$S_k \overset{\triangle}{=} \frac{1}{k}\left\{P - \left[(k-1)\frac{1}{\lambda_k} - \sum_{i \in \{1, \cdots, k-1\} \cap \{k \geq 2\}} \frac{1}{\lambda_i}\right]\right\}.$$

Now search for

$$k^* = \max\left\{k|S_k > 0, 1 \leq k \leq \hat{i}\right\}, \tag{3.6}$$

using the Fibonacci search method with approximation ratios $\frac{1}{3}$ and $\frac{2}{3}$. More specifically,

1st Step.   Assume that $a = 1$ and $b = \hat{i}$.
2nd Step.   If $a = b$, then $k^* = a$ and go to **Step 3**.
   Else, $a_1 = \lfloor a + \frac{1}{3}(b-a)\rfloor$, $b_1 = \lceil a + \frac{2}{3}(b-a)\rceil$.
3rd Step.   If $S_{a_1} \leq 0$, then $b = a_1 - 1$ and go to the **2nd Step**;
   If $S_{b_1} > 0$, then $a = b_1$ and go to the **2nd Step**;
   If $S_{a_1} > 0$ and $S_{b_1} \leq 0$, then $a = a_1, b = b_1 - 1$ and go to the **2nd Step**.

Step 3: Finding Optimal Solution to (3.5).

- Compute $S^* \in \mathbb{C}^{t \times t}$ as follows:

$$\begin{cases} S_{ii}^* = \frac{1}{\lambda_{k^*}} - \frac{1}{\lambda_i} + S_{k^*}, & \text{as } 1 \leq i \leq k^*; \\ S_{ii}^* = 0, & \text{as } k^* < i \leq t; \\ S_{ij}^* = 0, & \text{as } i \neq j. \end{cases} \tag{3.7}$$

- Compute $US^*U^\dagger$, as the optimal solution to the model (3.4). $US^*U^\dagger$ is proved to be the optimal solution in Chap. 2.

Remark 3.1. In Step 2 of the above GWF algorithm, the Fibonacci search method is used to find k^*. This can reduce the computational cost of computing

$$\max\left\{k|S_k > 0, 1 \leq k \leq \hat{i}\right\},$$

compared with the regular searching method of enumeration. Indeed, the ratio of the computation burden from the Fibonacci search to that from the enumeration method is about $\log(t)/t$. The impact of this reduction is amplified by the fact in iterative water-filling schemes, where the water-filling procedure is repeated many times. As anecdote, we point out that the water level obtained in the GWF procedure is $\frac{1}{\lambda_{k^*}} + S_{k^*}$.

3.1.3 Optimality and Complexity

For the channel capacity problem of the single user MIMO channel, the proof of optimality for $US^*U^\dagger$ found by the GWF algorithm with Fibonacci search is presented in this section.

Theorem 3.1. *Let $H = U\Sigma V^\dagger$ denote the singular value decomposition of the H. Further let k^* and S^* be given as in Step 2 and 3 of the GWF Algorithm with Fibonacci search; see (3.6) and (3.7). Then $US^*U^\dagger$ is an optimal input covariance for the problem in (3.4).*

Before proving this new theorem, we present some lemmas and introduce a remark.

Lemma 3.1. *For the channel H, there is a unitary matrix U such that $U^\dagger H^\dagger HU = \text{diag}(\lambda_1, \cdots, \lambda_t)$ (a diagonal matrix) and*

$$\max\left\{\log\left(\det\left(I_r + HSH^\dagger\right)\right) | S \succeq 0, \text{Tr}(S) \leq P\right\} =$$

$$\max\left\{\log\left(\det\left(I_t + \text{diag}(\lambda_1, \cdots, \lambda_t)S\right)\right) | S \succeq 0, \text{Tr}(S) \leq P\right\}$$

and $U^\dagger S_l U = S_r$, where S_l and S_r are two optimal solutions of the two optimization problems mentioned above, respectively.

Lemma 3.2. *For the channel H, there is a unitary matrix U such that $U^\dagger H^\dagger HU = \Lambda$ (a diagonal matrix) and*

$$\max\left\{\log\left(\det\left(I_r + HSH^\dagger\right)\right) | S \succeq 0, \text{Tr}(S) \leq P\right\}$$
$$= \max\left\{\log\left(\det\left(I_t + \Lambda^{\frac{1}{2}} S \Lambda^{\frac{1}{2}}\right)\right) | S \succeq 0, \text{Tr}(S) \leq P\right\},$$

and $U^\dagger S_l U = S_r$, where S_l and S_r are two optimal solutions of the two optimization problems mentioned above, respectively.

The proofs of both these lemmas are provided in Sect. A.3.

For simplification of the optimization model, which is

$$\max\left\{\log\left(\det\left(I_t + \Lambda^{\frac{1}{2}} S \Lambda^{\frac{1}{2}}\right)\right) | S \succeq 0, \text{Tr}(S) \leq P\right\},$$

the Hadamard Determinantal Inequality is introduced.

The Hadamard Determinantal Inequality [7] is: if $A = (a_{ij})_{n \times n}$ is a real (or Hermitian) positive semidefinite matrix, then

$$\det(A) \leq a_{11} \cdots a_{nn}.$$

Then we present our proof for the problem (3.5) as follows.

Proof of Theorem 3.1. When pre-processed, as the first step of the GWF algorithm,

$$\max\left\{\log\left(\prod_{i=1}^{t}(1+\lambda_i S_{ii})\right)\Big|S_{ii}\geq 0,\forall i,\sum_{i=1}^{t}S_{ii}\leq P\right\}$$

is equivalent to

$$\max\left\{\log\left(\prod_{i=1}^{\hat{i}}(1+\lambda_i S_{ii})\right)\Big|S_{ii}\geq 0,\forall i,\sum_{i=1}^{\hat{i}}S_{ii}= P\right\},$$

under the meaning of the equivalence for two optimization models.

$$\lambda_1\geq\lambda_2\geq\cdots\geq\lambda_{\hat{i}}\text{ and }\frac{1}{\lambda_1}\leq\frac{1}{\lambda_2}\leq\cdots\leq\frac{1}{\lambda_{\hat{i}}}.$$

Formulation

$$\max_{\{S_{ii}\}}\left\{\log\left(\prod_{i=1}^{\hat{i}}(1+\lambda_i S_{ii})\right)\Big|S_{ii}\geq 0,\forall i,\sum_{i=1}^{\hat{i}}S_{ii}= P\right\},\qquad(3.8)$$

is equivalent to (3.4) even although it has an equality constraint and the linear inequality constraints.

Below, similarly to proving Proposition 2.2, the set $\left\{S_{ii}^*,1\leq i\leq\hat{i}\right\}$ is proved to be the optimal solution to the problem in (3.8).

The Lagrangian function of the problem in (3.8) is

$$L\left(\{S_{ii}\}_{i=1}^{\hat{i}},\mu,\{\sigma_i\}_{i=1}^{\hat{i}}\right)= -\sum_{i=1}^{\hat{i}}\log\left(\frac{1}{\lambda_i}+S_{ii}\right)-\mu\left(P-\sum_{i=1}^{\hat{i}}S_{ii}\right)-\sum_{i=1}^{\hat{i}}\sigma_i S_{ii},$$

where μ and $\{\sigma_i\}_{i=1}^{\hat{i}}$ are the Lagrange multipliers. Therefore, the **KKT** conditions of the problem in (3.8) are

$$\begin{cases}\frac{1}{\frac{1}{\lambda_i}+S_{ii}}-\mu+\sigma_i=0, & 1\leq i\leq\hat{i}\\ S_{ii}\geq 0,\sigma_i S_{ii}=0,\ \sigma_i\geq 0,\forall i\\ \sum_{i=1}^{\hat{i}}S_{ii}=P, & \mu\in\mathbb{R}.\end{cases}\qquad(3.9)$$

Now we choose $S_{ii}=S_{ii}^*,1\leq i\leq\hat{i}$. Therefore,

$$\frac{1}{\frac{1}{\lambda_1}+S_{11}^*}=\cdots=\frac{1}{\frac{1}{\lambda_{k^*}}+S_{k^*k^*}^*}.$$

Now also choose $\mu = \frac{1}{\frac{1}{\lambda_1} + S_{11}^*}$ and choose $\sigma_i = 0, i = 1, \ldots, k^*$. By simple substitution, it can be shown that these values solve the **KKT** conditions.

To show that this point is optimal, we now argue that, for the problem in (3.8), the **KKT** conditions are sufficient as well as very necessary for optimality. First the Hessian matrix of the objective, $\log \left(\prod_{i=1}^{\hat{t}} (1 + \lambda_i S_{ii}) \right)$, is clearly negative definite and hence the objective is concave. Furthermore, the constraints are linear and hence the feasible set is convex.

Therefore, the problem in (3.8) is a convex optimization problem. Since only there are the linear constraints for the problem in (3.8), the constraint qualification is satisfied and the **KKT** conditions are both necessary and sufficient.

Since

$$\max \left\{ \log \left(\prod_{i=1}^{t} (1 + \lambda_i S_{ii}) \right) \Big| S_{ii} \geq 0, \forall i, \sum_{i=1}^{t} S_{ii} \leq P \right\}$$

is equivalent to

$$\max \left\{ \log \left(\prod_{i=1}^{\hat{t}} (1 + \lambda_i S_{ii}) \right) \Big| S_{ii} \geq 0, \forall i, \sum_{i=1}^{\hat{t}} S_{ii} = P \right\},$$

then $US^*U^\dagger$ is an optimal solution of the model (3.4).   **Q.E.D.**

Remark 3.2. Although there are other proofs available, the above proof was obtained independently and is simpler. In addition, it is easy to obtain computational complexity of the solution to the problem of the optimal input power under the single user MIMO case from Chap. 2.

3.2 Multi-user MIMO Channels

We consider two multiuser MIMO systems:

(i) The MIMO multiple access channel (MAC) in which a number of users wish to send messages to a single destination, and
(ii) The MIMO broadcast channel (BC) in which a single source wishes to send independent messages to different destinations.

These two classes are illustrated for the case of two users in the following diagram.

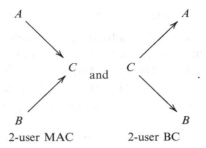

2-user MAC   2-user BC

The MAC model is used in the study of the cellular uplink (from user to base-station) and the BC model is used to study the cellular downlink.

In the case of the MAC and BC channels, the fundamental limit on the rates at which data can be reliably communicated is the capacity region. That is, the region of rate vectors for which there exists a coding strategy such that the probability of error goes to zero as the block length of the code increases. In this book, we will focus on one point on the boundary of the region, namely the maximum sum-rate point. That is, the point at which the sum of the rates is maximized. More specially, we will consider scenarios in which the channel coherence times are long, precise channel state information is available to both the transmitter and receivers, and the additive noise at the receivers is Gaussian. For this scenario, a popular algorithm for designing power allocation policy that maximizes the sum rate is the iterative water-filling algorithm (IWFA), such as the well known [5, 6]. In this book, two important contributions will be made to the generic IWFA. One contribution is to improve the convergence speed over the existing algorithms by using GWF. We also show that by taking a modified approach, rigorous convergence of the algorithms can be obtained for the MAC and BC models. These algorithms are based on our exploiting optimization theory and methods in communications. Since this exploiting must be implemented during IWFA, resulting efficiency can have a significant impact in practice.

For a class of the important problems that seeks to maximize the sum rate of the multi-user MIMO MAC and compute the corresponding optimal power allocation, we developed a more efficient algorithm for solving this class of the problems compared with currently known algorithms. The performance result of this new algorithm indicates that the proposed algorithm overcomes some of the weaknesses of other algorithms. One of the key weaknesses that it overcomes is that the well-known iterative water-filling algorithms cannot utilize the machinery of parallel computation, owning to their inherent structure defects. Not only does the proposed algorithm sufficiently utilize the machinery of parallel computation, it also shows faster convergence. Numerical results show that the same properties of the proposed algorithm are also effective for finding the optimal input policy due to its simplicity and fast convergence.

In our previous work [8], we have proven that IWFA is convergent to find the optimal power allocation policy of MIMO BC. However, although easily implemented, this algorithm has some limitations. One of the major limitations is that the greater the number of the users is, the slower convergence of the IWFAs appear to be. To address this limitation, this book first presents a new IWFA and shows its optimality result for the MIMO BC. When compared with previous research results, the performance of this proposed algorithm has a strong robustness with respect to the number of the users/channels; In addition, parallel processing can be utilized to expedite the speed of the computation. Therefore, the proposed new algorithm is effective for finding the optimal transmission policy due to its simplicity and fast convergence.

3.3   Efficient RRM of the Multi-user MIMO MAC Channels

MIMO MAC is a promising technology for the next generation of wireless communication systems. For a MIMO MAC system, it is important to find the optimal input (power) distribution which optimizes the resource allocation policies of the Gaussian MIMO MAC with multiple antennas at the base station and each of the users. The optimal input distribution needs to be found by efficient algorithms.

As one of the fundamental algorithms to compute the optimal input distribution which maximizes the sum rate, i.e., sum capacity, for Gaussian MIMO MAC, an IWFA to find the maximum of the sum rate and the corresponding optimal input distribution was presented in [6]. This algorithm has also been employed to solve the problem of optimal power allocation over frequencies of the scalar intersymbol interference (ISI) multiple-access channels and that of the scalar independent and identically distributed ($i.i.d$) fading channels. In [6], the ISI channels and the $i.i.d$ fading channels mentioned above are considered as only two special cases of the sum rate maximization problem of the MIMO MAC. In addition, the algorithm of [6] has also been extended in [9].

However, the current form of the IWFA has limitations when it comes to efficiently find the sum capacity of the MIMO MAC. One of the major limitations is that it cannot utilize the machinery of parallel computation owing to its inherent structure defects. Due to the ongoing trend of employing multi-core processing for efficiency purpose, the machinery of parallel computation and pipeline are important from the algorithm point of view. Hence, it is an important challenging task to design an algorithm not only shows faster convergence, but it also sufficiently utilizes the machinery of parallel computation. To meet these challenges for the general model of the MIMO MAC, this chapter presents a new and effective convergent algorithm based on the algorithm of [6] with the aforementioned characteristics as its own unique advantages.

The rest of this section is organized as follows. In Sect. 3.3.1, the proposed algorithm: iterative geometric water-filling for MAC cases (IGWF-MAC) is presented for the model of the sum capacity of Gaussian MIMO MAC under the individual power constraints. In Sect. 3.3.2, its convergence is discussed.

3.3.1 Models for the MIMO MAC and Its Sum Capacities

MIMO is a promising technology for the next generation wireless communication systems. Multi-user MIMO (MU-MIMO) technologies exploit the availability of multiple independent radio terminals in order to enhance the communication capabilities of the system. Being equipped with multiple antennas at the base station and at each of the mobile stations, the MU-MIMO may be generalized into two categories: MIMO BC and MIMO MAC for downlink and uplink situations, respectively. Then, it is important to find the optimal input power distribution which optimizes the resource allocation policies. In this section, we focus our discussion on the optimal resource allocation for MIMO MAC. Due to complicated structures of these resource allocation policy, the optimized solutions might not be computed directly. Instead, an algorithmic approach, i.e. iterative algorithm, is used to compute the optimal input power distribution for the MIMO MAC.

For the MIMO MAC case, we assume that the base station has m antennas, and there are K mobile stations, each of which has s antennas. When the information flows from the mobile stations to the base station, we have a MAC channel or uplink channel.

With appropriate synchronization, the MAC can be described as [4]

$$y_{\text{mac}} = \sum_{i=1}^{K} H_i^\dagger x^i + z, \tag{3.10}$$

where y_{mac} is the signal received at the base-station, and $H_i^\dagger \in \mathbb{C}^{m \times s}, i = 1, 2, \cdots, K$, denotes the matrix of channel gains from each antenna at the i-th mobile station to each antenna at the base station. Without loss of generality, we will assume that $H_i \neq 0$, $\forall i$. The x^is are $s \times 1$ complex input vectors, and $z \in \mathbb{C}^m$ is an additive Gaussian noise vector and, without loss of generality, has identity covariance. We will let $S_i \stackrel{\triangle}{=} E\left[x^i \left(x^i\right)^\dagger\right]$ denote the covariance matrix of x^i. For the convenience of later discussion and without loss of generality, the MAC is described as $y_{\text{mac}} = \sum_{i=1}^{K} H_i^\dagger x^i + z$ [4] instead of $y_{\text{mac}} = \sum_{i=1}^{K} H_i x^i + z$. Notation-wise, in this section, we will use different superscripts of vectors to denote vectors corresponding to different users and different subscripts of a vector to denote different entries of a vector.

The sum capacity of the MIMO MAC is the fundamental limit on the sum of the rates at which reliable communication can be achieved. When the channels are known and deterministic and the noise is Gaussian, the optimal input distribution at each mobile station is a zero-mean Gaussian random vector with covariance S_i. Therefore, the problem of maximizing the sum rate, or the sum capacity, can be written as follows [4].

$$C_{mac}\left(H_1^\dagger, \cdots, H_K^\dagger, P\right) = \max_{\{S_i\}_{i=1}^{K}: S_i \succeq 0, \text{Tr}(S_i) \leq P_i} \log \left|I + \sum_{i=1}^{K} H_i^\dagger S_i H_i\right|, \tag{3.11}$$

where $S_i \succeq 0$ denotes that S_i is a Hermitian positive semidefinite matrix with complex entries.

Without loss of generality, we assume quad-core parallel computation is to be utilized. Prior to introducing the proposed algorithm, Algorithm IGWF-MAC, some concepts used in the algorithm are presented as below.

Let

$$\kappa^{(n)} = \{[4n+1]_K, [4n+2]_K, [4n+3]_K, [4n+4]_K\}, \tag{3.12}$$

where $[x]_K = \mod((x-1), K) + 1$ and $n = 0, 1, \ldots$ For $\forall \kappa^{(n)}$ and given feasible point of the optimization problem (3.11), $\left\{ S_i^{(n)} \right\}_{i=1}^{K}$, we denote

$$\left\{ S_{[4n+1]_K}^*, S_{[4n+2]_K}^*, S_{[4n+3]_K}^*, S_{[4n+4]_K}^* \right\} \tag{3.13}$$

by

$$\left\{ S_{[4n+j]_K}^* \right\}_{j=1}^{4} \stackrel{\triangle}{=} \arg\max_{\{S_k\}_{k\in\kappa^{(n)}} : S_k \succeq 0, \mathrm{Tr}(S_k) \leq P_k} \\ \sum_{k\in\kappa^{(n)}} \log |H_k^\dagger S_k H_k + \sum_{1\leq j\leq K, j\neq k} H_j^\dagger S_j^{(n)} H_j + I|. \tag{3.14}$$

Note that to solve this optimization problem, the quad-core parallel computation corresponds to the four-element set $\kappa^{(n)}$. We denote by f_1 the objective function of (3.14). The updated covariance matrices are defined by

$$\overline{S}_{k,t} = t S_k^* + (1-t) S_k^{(n)}, \text{ as } k \in \kappa^{(n)} \tag{3.15}$$

and $t = \frac{1}{4}, \frac{2}{4}, \frac{3}{4}, 1$. Also let us define a function by

$$f_0\left(\{S_k\}_{k\in\kappa^{(n)}}\right) = \log|I + \sum_{k\in\kappa^{(n)}} H_k^\dagger S_k H_k + \sum_{1\leq l\leq K, l\notin\kappa^{(n)}} H_l^\dagger S_l^{(n)} H_l| \tag{3.16}$$

Another covariance or power update step is defined as:

$$\begin{cases} S_k^{(n+1)} = S_k^{(n)}, & \text{if } k \notin \kappa^{(n)}; \\ S_k^{(n+1)} = \overline{S}_{k,t^*}, & \text{if } k \in \kappa^{(n)}, \end{cases} \tag{3.17}$$

where $t^* = \max\arg\max_{t\in\{\frac{i}{4}\}_{i=1}^{4}} f_0\left(\{\overline{S}_{k,t}\}_{k\in\kappa^{(n)}}\right)$. $n = n+1$ and then return to (3.14).

According to (3.17), the covariance update steps can be denoted by the mapping

$$f: \left\{S_k^{(n+1)}\right\}_{k=1}^{K} = f\left(\left\{S_k^{(n)}\right\}_{k=1}^{K}\right).$$

With the assumptions and the concepts introduced, a new iterative water-filling algorithm, IGWF-MAC, is concisely proposed as follows.

Algorithm IGWF-MAC:

1. Initialize $\left\{S_k^{(0)}\right\}_{k=1}^K = 0$.
2. Compute $\left\{S_k^{(n+1)}\right\}_{k=1}^K = f\left(\left\{S_k^{(n)}\right\}_{k=1}^K\right)$. Then $n = n+1$.
3. Repeat the procedure (2) mentioned above until the sum capacity converges.

Remark 3.3. For implementing Step 2 in IGWF-MAC, (3.14) and (3.17) are utilized. Their details are stated below.

1. Generate effective channels

$$G_k^{(n)} = H_k \left(I + \sum_{j\neq k} H_j^\dagger S_j^{(n)} H_j\right)^{-\frac{1}{2}}, \; k \in \kappa^{(n)}. \tag{3.18}$$

2. Treating these effective channels as parallel and noninterfering channels, the new covariance matrices $\left\{\overline{S}_{k,t}\right\}_{k\in\kappa^{(n)}}$ are generated by GWF with Fibonacci search (to be defined later) under the individual power constraints, i.e., $\text{Tr}(S_k) \leq P_k$ with $k \in \kappa^{(n)}$, where $t = \frac{1}{4}, \frac{2}{4}, \frac{3}{4}, 1$.

$$\left\{\overline{S}_{k,t}\right\}_{k\in\kappa^{(n)}} = \arg \max_{\{S_k\}_{k\in\kappa^{(n)}} : S_k \succeq 0, \text{Tr}(S_k)\leq P_k} \sum_{k\in\kappa^{(n)}} |I + \left(G_k^{(n)}\right)^\dagger S_k G_k^{(n)}|, \tag{3.19}$$

3. Update step: Due to

$$t^* = \max\{\arg \max_{t\in\{\frac{i}{4}\}_{i=1}^4} f_0\left(\left\{\overline{S}_{k,t}\right\}_{k\in\kappa^{(n)}}\right)\},$$

and the function f_0 having been defined above, we may know that f_0 is concave. Hence, the golden section principle [10] is used for searching t^*. Its detail is provided as follows.

1st Step.
    Assume that $a = 1/4, b = 1, a_1 = a + \frac{1}{4}, b_1 = b - \frac{1}{4}$.
2nd Step.
    If $f(a_1) \leq f(b_1)$, then $a = b_1$. Further, for next step, if $f(a) \leq f(b)$, then $t^* = b$; Else, $t^* = a$.
    If $f(a_1) > f(b_1)$, then $t^* = a_1$.

Convergence and fast convergence of the algorithm are to be proved and illustrated in later subsections in detail.

To carry out the GWF step, as Step 2 of Algorithm IGWF-MAC, we may choose to use CWF, which is an enumeration method. However since the water-filling algorithm is utilized repeatedly, reducing its computation amount is particularly meaningful. Hence in Step 2 of Algorithm IGWF-MAC, we design such an algorithm by the mean of Fibonacci search and construct a set of explicit solution to the optimization problem (3.27). We can reduce the computation amount by a factor of at least 1/3 this way when compared with the enumeration method. Our method, called the iterative water-filling algorithm with Fibonacci search (IWFAwFIS) for IGWF-MAC, is stated in the following three steps as the solution to the problem (3.27).

Step 1: Pre-Processing.

For $k \in \kappa^{(n)}$, compute the unitary matrix $U_k^{(n)} \in \mathbb{C}^{s \times s}$ by calculating the eigenvalue decomposition satisfying

$$
\begin{aligned}
\Lambda_k^{(n)} &= \left(U_k^{(n)} \right)^\dagger G_k^{(n)} \left(G_k^{(n)} \right)^\dagger U_k^{(n)} \\
&= \mathrm{diag} \left(\left(\Lambda_k^{(n)} \right)_{1,1}, \cdots, \left(\Lambda_k^{(n)} \right)_{s,s} \right).
\end{aligned}
\tag{3.20}
$$

Let $\left\{ \left(\Lambda_k^{(n)} \right)_{1,1}, \cdots, \left(\Lambda_k^{(n)} \right)_{s,s} \right\}$ be ordered monotonically decreasing into $\{\lambda(k)_t\}_{t=1}^s$.

$$
j(k) \triangleq \max \left\{ j | \lambda(k)_j > 0 \right\}.
$$

Step 2: GWF with Fibonacci Search.

Let

$$
R(k)_j \triangleq \frac{1}{j} \left\{ P_k - \left[(j-1) \frac{1}{\lambda(k)_j} - \sum_{t=1}^{j-1} \frac{1}{\lambda(k)_t} \right] \right\}.
$$

Search

$$
k^* = \max \left\{ j | R(k)_j > 0, 1 \leq k \leq j(k) \right\}.
$$

Here, the Fibonacci approximation ratio $\frac{1}{3}$ and $\frac{2}{3}$ are used for searching k^*, and this method is called the Fibonacci search. The searching of k^* is presented as follows.

1st Step.

Assume that $a = 1$ and $b = j(k)$.

2nd Step.

If $a = b$, then $k^* = a$ and go to Step 3.

Else, $a_1 = \lfloor a + \frac{1}{3}(b-a) \rfloor$, $b_1 = \lceil a + \frac{2}{3}(b-a) \rceil$. Note that "$\lfloor \ \rfloor$" stands for the floor function, and "$\lceil \ \rceil$" stands for the ceiling function.

3rd Step.
   If $R(k)_{a_1} \leq 0$, then $b = a_1 - 1$ and go to the 2nd step;
   If $R(k)_{b_1} > 0$, then $a = b_1$ and go to the 2nd step;
   If $R(k)_{a_1} > 0$ and $R(k)_{b_1} \leq 0$, then $a = a_1$, $b = b_1 - 1$ and go to the 2nd step.

Step 3: Find Optimal Solution of (3.27).

1st Step.
   Compute $S_k^* \in \mathbb{C}^{s \times s}$ as follows:

$$
\begin{cases}
\left(S_k^*\right)_{t,t} = \frac{1}{\lambda(k)_{k^*}} - \frac{1}{\left(\Lambda_k^{(n)}\right)_{t,t}} + R(k)_{k^*}, \text{ where } 1 \leq t \leq k^*; \\
\left(S_k^*\right)_{t,t} = 0, \; k^* < t \leq s; \\
\left(S_k^*\right)_{l,t} = 0, \; l \neq t.
\end{cases}
\tag{3.21}
$$

2nd Step.
   Compute $S_k^{(n+1)} = U_k^{(n)} S_k^* \left(U_k^{(n)}\right)^{\dagger}$.

In the next section, the optimality of $\left\{S_k^{(n+1)}\right\}_{k=1}^K$ is to be proved, i.e., $\left\{S_k^{(n+1)}\right\}_{k \in \kappa^{(n)}}$ is the solution to (3.27).

Remark 3.4. Based on the structure of the model (3.27) in Algorithm IGWF-MAC, parallel processing can be utilized for more efficient computations.

Remark 3.5. The golden section principle in Step 2 of Algorithm IGWF-MAC is utilized twice. For the first time, it is used to reduce the computation burden from the water-filling of (3.14). For the second time, it is used efficiently to speed up convergence of the algorithm and to improve performance of the algorithm, although it may seem to only compute t^* quickly. This point will be observed from the numerical experiments in Sect. 3.5.

Remark 3.6. IGWF-MAC is a simple and effective algorithm due to its corresponding objective function $V\left(\beta \gamma^{(n)} + (1 - \beta) p^{(n-1)}\right)$ being concave in the scalar variable β. In fact, we can also choose any finite searching steps of the function $V\left(\beta \gamma^{(n)} + (1 - \beta) p^{(n-1)}\right)$ to improve the performance of the algorithm. Even with these modifications, we can still guarantee convergence of Algorithm IGWF-MAC with the previously mentioned advantages. The focus of this section is on the improvement of the iterative water-filling algorithm of [6]. More specifically, Steps 2 and 3 of Algorithm IGWF-MAC permit us to employ the Fibonacci search principle and machinery of parallel computation. This in turn achieves much faster convergence, with almost finite-step convergence.

3.3.2 Optimality of Power Allocation Algorithm: IGWF-MAC

First, we prove the optimality of $\left\{S_k^{(n+1)}\right\}_{k\in\kappa^{(n)}}$, which is found by Step 3 of IWFAwFIS for IGWF-MAC, as a set of solution to (3.27). Then the continuous mapping is introduced, and according to the concept of the closed mapping and its relationship to a continuous mapping being closed, convergence of Algorithm IGWF-MAC can be formally proved.

Proposition 3.1. *For the problem (3.27),* $\left\{S_k^{(n+1)}\right\}_{k\in\kappa^{(n)}}$, *obtained by Step 3 of IWFAwFIS for IGWF-MAC, is the optimal solution to the problem (3.27).*

Proof. It is easily known that the optimization problem (3.27) is equivalent to

$$\max_{\{\tilde{S}_k\}_{k\in\kappa^{(n)}}:\tilde{S}_k\succeq 0,\mathrm{Tr}(\tilde{S}_k)\le P_k}\sum_{k\in\kappa^{(n)}}\log\left|I+(\Lambda_k^{(n)})^{\frac{1}{2}}\tilde{S}_k(\Lambda_k^{(n)})^{\frac{1}{2}}\right|, \qquad (3.22)$$

where there is a unitary matrix $U_k^{(n)}$ such that

$$\left(U_k^{(n)}\right)^{\dagger}G_k^{(n)}\left(G_k^{(n)}\right)^{\dagger}U_k^{(k)}=\Lambda_k^{(n)}=\mathrm{diag}\left(\left(\Lambda_k^{(n)}\right)_{1,1},\cdots,\left(\Lambda_k^{(n)}\right)_{s,s}\right),\forall k. \qquad (3.23)$$

Let $\left\{\left(\Lambda_k^{(n)}\right)_{1,1},\cdots,\left(\Lambda_k^{(n)}\right)_{s,s}\right\}$ be ordered monotonically decreasing into $\{\lambda(k)_j\}_{j=1}^{s}$. Further, the optimization problem (3.31) is equivalent to a set of the following problems:

$$\begin{aligned}\max_{\left\{(\tilde{S}_k)_{t,t}\right\}}&\sum_{t=1}^{s}\log\left(1+\left(\Lambda_k^{(n)}\right)_{t,t}(\tilde{S}_k)_{t,t}\right)\\\text{subject to }&\sum_{t=1}^{s}(\tilde{S}_k)_{t,t}\le P_k;\\&(\tilde{S}_k)_{l,t}=0,\\&\text{as }l\ne t,1\le l,t\le s;\ k\in\kappa^{(n)}.\end{aligned} \qquad (3.24)$$

Given $\overline{S}_k^{(n)}=U_k^{(n)}\overline{S}_k^{*}\left(U_k^{(n)}\right)^{\dagger}$, we will prove that $\left\{\overline{S}_k^{(n)}\right\}_{k\in\kappa^{(n)}}$ and $\left\{\overline{S}_k^{*}\right\}_{k\in\kappa^{(n)}}$ are the optimal solutions of the optimization problems (3.27) and (3.33), respectively.

According to the first step in Step 3 of IWFAwFIS for IGWF-MAC, the set $\left\{(S_k^{*})_{t,t},1\le t\le k^{*}\right\},\forall k$, implies

$$\frac{1}{\frac{1}{\left(\Lambda_k^{(n)}\right)_{t,t}}+(S_k^{*})_{t,t}}=\frac{1}{\frac{1}{\lambda(k)_{k^{*}}}+R(k)_{k^{*}}},1\le t\le k^{*},\forall k.$$

There is a number μ_k which satisfies the following equalities.

$$\mu_k = \frac{1}{\frac{1}{\left(\Lambda_k^{(n)}\right)_{t,t}} + \left(S_k^*\right)_{t,t}} = \frac{1}{\frac{1}{\lambda(k)_{k^*}} + R(k)_{k^*}}, 1 \leq t \leq k^*.$$

Let the number $\sigma_{k,t} = 0, t = 1, \ldots, k^*, \forall k$. Due to, as $k^* < t \leq s$,

$$\mu_k = \frac{1}{\frac{1}{\lambda(k)_{k^*}} + R(k)_{k^*}} \geq \left(\Lambda_k^{(n)}\right)_{t,t},$$

$$\sigma_{k,t} = \mu_k - \left(\Lambda_k^{(n)}\right)_{t,t} \geq 0.$$

For any k, the corresponding Lagrange function of the problems (3.33) is

$$
\begin{aligned}
L\left(\left\{\left(\tilde{S}_k\right)_{t,t}\right\}; \overline{\mu}_k, \left\{\overline{\sigma}_{k,t}\right\}\right) &= \Sigma_{t=1}^s \log\left(1 + \left(\Lambda_k^{(n)}\right)_{t,t}\left(\tilde{S}_k\right)_{t,t}\right) \\
&\quad + \overline{\mu}_k\left(P_k - \Sigma_{t=1}^s \left(\tilde{S}_k\right)_{t,t}\right) \\
&\quad + \Sigma_{t=1}^s \overline{\sigma}_{k,t}\left(\tilde{S}_k\right)_{t,t},
\end{aligned}
\tag{3.25}
$$

where $\overline{\mu}_k$ and $\left\{\overline{\sigma}_{k,t}\right\}$ are the Lagrange multipliers. It is easily known that the set $\left\{(S_k^*)_{t,t}\right\}$, μ_k and $\left\{\sigma_{k,t}\right\}$ mentioned above satisfy the Karush-Kuhn-Tucker (KKT) conditions of the equivalent model, i.e.,

$$
\begin{cases}
\frac{1}{\frac{1}{\left(\Lambda_k^{(n)}\right)_{t,t}} + \left(S_k^*\right)_{t,t}} - \mu_k + \sigma_{k,t} = 0, & 1 \leq t \leq s; \\
\left(S_k^*\right)_{t,t} \geq 0, & \\
\sigma_{k,t}\left(S_k^*\right)_{t,t} = 0, & \\
\sigma_{k,t} \geq 0, & \forall k,t; \\
\Sigma_{t=1}^s \left(S_k^*\right)_{t,t} = P_k, & \mu_k \in \mathbb{R}.
\end{cases}
$$

Due to each of the problems (3.33) being a convex optimization problem with a concave objective function, $\left\{S_k^*\right\}_{k \in \kappa^{(n)}}$ is a set of the optimal solutions to the problems (3.33), respectively. Due to the equivalence between the problem (3.27) and the problems (3.33), $\left\{\overline{S}_k^{(n)}\right\}_{k \in \kappa^{(n)}}$, obtained by Step 3 of IWFAwFIS for IGWF-MAC, is indeed the optimal solution to the problem (3.27).

Second, we apply the fixed point theory set up by us with a series of derivations, the convergence for the algorithm IGWF-MAC is then obtained. To reduce the cost of computation, (2) in Remark 3.3 of IGWF-MAC utilizes Fibonacci search. To improve the performance of the algorithm and reduce the cost of computation for the sum rate optimization problem mentioned above, the objective function in (3) of

Remark 3.3 for IGWF-MAC may only be evaluated at the four points of $t = \frac{1}{4}, \frac{2}{4}, \frac{3}{4}$ and 1, corresponding to (3.17), by parallel computation to find t^*. The following numerical experiments show the improvement of the algorithm.

3.4 Efficient RRM of the Multi-user MIMO BC Channels

For a MIMO BC system, it is important to find the optimal transmission distribution which optimizes the resource allocation policies of the Gaussian MIMO BC with multiple antennas at the base station. The optimal transmission distribution needs to be found by efficient algorithms.

In order to compute the optimal transmission distribution, the maximization of the sum-rate for Gaussian MIMO BC, i.e., sum capacity under the constraint of the sum power, has been investigated in many papers, such as [5,9,11,12]. For the general model of the MIMO BC, on one hand, the proposed algorithm provides stronger robustness on the number of the users when compared with the work of [11], and at the same time, it offers simplicity from the algorithm structure perspective and fast convergence with the iterative operations; On the other hand, convergence of the designed algorithm can be also guaranteed through rigorous mathematical proof. To meet these challenges, we present a new and effective convergent algorithm with the aforementioned characteristics as its unique advantages. The second point mentioned, i.e., convergence of the designed algorithm, seems to lay down a more strict fundamental basis as an improvement of others.

3.4.1 Models for the MIMO BC and Its Sum Capacities

For a downlink channel or the BC (see [4]), there are one base station with m antennas and K mobile stations each of which has n antennas. In this section, assume that the downlink channel is described as $y_i = H_i x + Z_i$, where $H_i \in \mathbb{C}^{n \times m}, i = 1, 2, \cdots, K$, are the fixed channel matrices and Z_i is an additive Gaussian noise with unit variance. The dual uplink channel of the downlink channel can be described as $y_{dmac} = \sum_{i=1}^{K} H_i^\dagger x^i + Z$, which is the same as that of MIMO MAC (see [13]). If $S_i \overset{\triangle}{=} E\left[x^i \left(x^i\right)^\dagger\right]$, then the mathematical model [4,13,14] of the sum capacity of the dual MIMO MAC is:

$$C_{dmac}\left(H_1^\dagger, \cdots, H_K^\dagger, P\right) =$$
$$\max_{\{S_i\}_{i=1}^{K}: S_i \succeq 0, \sum_{i=1}^{K} \mathrm{Tr}(S_i) \leq P} \log\left|I + \sum_{i=1}^{K} H_i^\dagger S_i H_i\right|. \tag{3.26}$$

This mathematical model (or problem) of the sum capacity of the dual MIMO MAC is in fact the sum rate optimization problem of the MIMO BC. The constraints of the sum rate optimization problem of the MIMO BC are called the sum power constraint. By the way, the constraints of the sum rate optimization problem of the MIMO MAC are the individual power constraints. Compared with the optimization problem (3.26), the sum rate optimization problem of the MIMO MAC is simpler. There are two reasons for introducing this dual. The first is that the sum capacity of the MIMO BC is equal to that of the dual MIMO MAC, i.e., $C_{BC}(H_1, \cdots, H_K, P) = C_{dmac}\left(H_1^\dagger, \cdots, H_K^\dagger, P\right)$; and the second is that the latter can be much more easily determined than the former.

Therefore, for easily finding the sum capacity of the MIMO BC, only the dual MIMO MAC, with the equivalence of the optimal values, i.e., $C_{BC}(H_1, \cdots, H_K, P) = C_{dmac}\left(H_1^\dagger, \cdots, H_K^\dagger, P\right)$, is being considered here [4]. Note that the power constraint *couples* the stages for the sum rate optimization problem of the MIMO BC. For concision, the objective function $C_{dmac}\left(H_1^\dagger, \cdots, H_K^\dagger, P\right)$ in (3.26) is also denoted by symbol f. For effectively finding the sum capacity of the MIMO BC, according to the relationship, i.e., $C_{dmac}\left(H_1^\dagger, \cdots, H_K^\dagger, P\right) = C_{BC}(H_1, \cdots, H_K, P)$, a new algorithm is presented below.

Algorithm: Iterative Geometric Water-Filling (IGWF-BC):

Input: matrix $H_i, Q_i^{(0)} = 0$, $i = 1, \ldots, K$.

1. Generate effective channels

$$G_i^{(\tilde{n})} = H_i \left(I + \sum_{j \neq i} H_j^\dagger Q_j^{(\tilde{n}-1)} H_j\right)^{-\frac{1}{2}}, \text{ for } i = 1, \ldots, K.$$

2. Treating these effective channels as parallel, noninterfering channels, the new covariance matrices $\left\{Q_i^{(\tilde{n})}\right\}_{i=1}^K$ is generated by GWF with Fibonacci search (to be defined later) under the sum power P constraint,

$$\left\{Q_i^{(\tilde{n})}\right\}_{i=1}^K = \arg \max_{\{Q_i\}_{i=1}^K : Q_i \succeq 0, \sum_{i=1}^K \text{Tr}(Q_i) \leq P} \sum_{i=1}^K \log\left|I + \left(G_i^{(\tilde{n})}\right)^\dagger Q_i G_i^{(\tilde{n})}\right|, \quad (3.27)$$

3. Update step: Let

$$\beta^* = \max\left\{\beta_1 \in \arg \max_{\beta \in [1/K, 1]} f\left(\beta \gamma^{(\tilde{n})} + (1-\beta) p^{(\tilde{n}-1)}\right)\right\},$$

and

$$p^{(\tilde{n})} = \beta^* \gamma^{(\tilde{n})} + (1 - \beta^*) p^{(\tilde{n}-1)} \quad (3.28)$$

where the function f has been defined in (3.26), and $\gamma^{(\tilde{n})} \triangleq \left(Q_1^{(\tilde{n})}, Q_2^{(\tilde{n})}, \cdots, Q_K^{(\tilde{n})} \right)$ and $p^{(\tilde{n}-1)} \triangleq \left(Q_1^{(\tilde{n}-1)}, Q_2^{(\tilde{n}-1)}, \cdots, Q_K^{(\tilde{n}-1)} \right)$. Note that (3.28) means that we assign $\beta^* \gamma^{(\tilde{n})} + (1 - \beta^*) p^{(\tilde{n}-1)}$ to $p^{(\tilde{n})}$, and then $p^{(\tilde{n})}$ may be labeled into

$$\left(Q_1^{(\tilde{n})}, Q_2^{(\tilde{n})}, \cdots, Q_K^{(\tilde{n})} \right).$$

Here, the golden section method is used for searching β^*. Its detail is provided as follows.

1st Step.
  Let $\delta > 0$ as a permitted error. Assume that $a = 1/K$ and $b = 1$.
2nd Step.
  If $|a - b| \leq \delta$, then $\beta^* = \frac{1}{2}(a + b)$ and output β^*.
  Else, $a_1 = a + 0.382(b - a)$, $b_1 = a + 0.618(b - a)$.
3rd Step.
  If $f(a_1) < f(b_1)$, then $a = a_1$ and go to the 2nd step;
  If $f(a_1) > f(b_1)$, then $b = b_1$ and go to the 2nd step;
  If $f(a_1) = f(b_1)$, then $a = b_1$ and go to the 2nd step.

$\tilde{n} = \tilde{n} - 1$.

Go to 1.

This new algorithm has the same advantages that were previously claimed by Kobayashi and Caire [11], i.e., not being sensitive to the number of the users K, the simplicity of the algorithmic structure, the guaranteed convergence, and fast convergence of the algorithm despite large K. Convergence and fast convergence of the algorithm are to be proved and illustrated in later subsections.

To implement the GWF step, as (2) of Algorithm IGWF-BC, we may choose to use CWF, which is an enumeration method. However since the water-filling algorithm is utilized repeatedly, reducing its computation amount is particularly meaningful. Hence in (2) of Algorithm IGWF-BC, we design such an algorithm by the mean of Fibonacci search and construct a set of explicit solution to the optimization problem (3.27). We can reduce the computation amount by a factor of at least 1/3 this way when compared with the enumeration method for large K. Our method, called the iterative water-filling algorithm with Fibonacci search (IWFAwFIS) for IGWF-BC, is stated in the following three steps as the solution to the problem (3.27).

Step 1: Pre-Processing.
  Compute the unitary matrix $U_i^{(\tilde{n})} \in \mathbb{C}^{n \times n}$ by calculating the eigenvalue decomposition satisfying

$$\Lambda_i = \left(U_i^{(\tilde{n})} \right)^\dagger G_i^{(\tilde{n})} \left(G_i^{(\tilde{n})} \right)^\dagger U_i^{(\tilde{n})}$$
$$= \operatorname{diag} \left((\Lambda_i)_{1,1}, \cdots, (\Lambda_i)_{n,n} \right). \tag{3.29}$$

Let $\left\{ (\Lambda_1)_{1,1}, \cdots, (\Lambda_K)_{n,n} \right\}$ be ordered monotonically decreasing into $\{\lambda_t\}_{t=1}^{K \times n}$.

$$j(i) \overset{\triangle}{=} \max \left\{ j | (\Lambda_i)_{j,j} > 0 \right\}$$

and let

$$\bar{n} \overset{\triangle}{=} \sum_{i=1}^{K} j(i) \leq \min \{Kn, Km\}.$$

Step 2: GWF with Fibonacci Search.
Let

$$S_k \overset{\triangle}{=} \frac{1}{k} \left\{ P - \left[(k-1) \frac{1}{\lambda_k} - \sum_{t=1}^{k-1} \frac{1}{\lambda_t} \right] \right\}.$$

Search

$$k^* = \max \{ k | S_k > 0, 1 \leq k \leq \bar{n} \}$$

and

$$k_i^* = \max \left\{ j | (\Lambda_i)_{j,j} = \lambda_k, S_k > 0, 1 \leq k \leq \bar{n} \right\}.$$

Here, the Fibonacci approximation ratio $\frac{1}{3}$ and $\frac{2}{3}$ are used for searching k^* and k_i^*, and this method is called the Fibonacci search. To avoid repeating, the searching of k^* may be seen in [1] for the detail. The searching of k_i^* is presented as follows.

1st Step.
  Assume that $a = 1$ and $b = j(i), \forall i$.
2nd Step.
  If $a = b$, then $k_i^* = a$ and go to Step 3.
  Else, $a_1 = \lfloor a + \frac{1}{3}(b-a) \rfloor$, $b_1 = \lceil a + \frac{2}{3}(b-a) \rceil$, where "$\lfloor \ \rfloor$" stands for the floor function, and "$\lceil \ \rceil$" stands for the ceiling function.
3rd Step.
  If $(\Lambda_i)_{a_1,a_1} < \lambda_{k^*}$, then $b = a_1 - 1$ and go to the 2nd step;
  If $(\Lambda_i)_{b_1,b_1} \geq \lambda_{k^*}$, then $a = b_1$ and go to the 2nd step;
  If $(\Lambda_i)_{a_1,a_1} \geq \lambda_{k^*} > (\Lambda_i)_{b_1,b_1}$, then $a = a_1, b = b_1 - 1$ and go to the 2nd step.

Step 3: Find Optimal Solution of (3.27).

1st Step.
   Compute $S^* \in \mathbb{C}^{n \times n}$ as follows:

$$
\begin{aligned}
(S_i^*)_{t,t} &= \frac{1}{\lambda_{k^*}} - \frac{1}{(\Lambda_i)_{t,t}} + S_{k^*}, \\
&\text{where } 1 \le t \le k_i^*; \\
(S_i^*)_{t,t} &= 0, k_i^* < t \le n; \\
(S_i^*)_{s,t} &= 0, s \ne t.
\end{aligned}
\tag{3.30}
$$

2nd Step.
   Compute $Q_i^{(\tilde{n})} = U_i^{(\tilde{n})} S_i^* \left(U_i^{(\tilde{n})} \right)^{\dagger}$.

In the next subsection, the optimality of $\left\{ Q_i^{(\tilde{n})} \right\}_{i=1}^{K}$ is to be proved, i.e., $\left\{ Q_i^{(\tilde{n})} \right\}_{i=1}^{K}$ is the solution to (3.27).

Remark 3.7. Based on the structure of the model (3.27) in Algorithm IGWF-BC, parallel processing can be utilized for more efficient computations.

Remark 3.8. The golden section method in step 3 of Algorithm IGWF-BC is a simple and effective method due to its corresponding objective function

$$
f \left(\beta \gamma^{(n)} + (1 - \beta) p^{(n-1)} \right)
$$

being convex in the scalar variable β, compared with the evaluation of the objective function of the model (3.27). In fact, we can also choose any finite searching steps with even fewer evaluations of the function

$$
f \left(\beta \gamma^{(n)} + (1 - \beta) p^{(n-1)} \right).
$$

Even with these modifications, we can still guarantee convergence of Algorithm IGWF-BC with the previously mentioned advantages. The focus of this chapter is on the improvement of the Algorithm 2 of [5]. More specifically, Step 3 of Algorithm IGWF-BC employs a larger weight for the innovation when compared with that of [5]. This in turn achieves much faster convergence, and almost insensitive to the value of K.

3.4.2 Optimality of Power Allocation Algorithm: IGWF-BC

First, we prove the optimality of $\left\{ Q_i^{(\tilde{n})} \right\}_{i=1}^{K}$, which is found by the step 3 of IWFAwFIS for IGWF-BC, as a set of solution to (3.27). Then the continuous mapping is introduced, and according to the concept of the closed mapping and its relationship to a continuous mapping being closed, convergence of Algorithm IGWF-BC can be formally proved.

Proposition 3.2. *For the problem (3.27), $\left\{Q_i^{(\tilde{n})}\right\}_{i=1}^{K}$, obtained by the step 3 of IWFAwFIS for IGWF-BC, is the optimal solution to the problem (3.27).*

Proof. It is easily known that the optimization problem (3.27) is equivalent to

$$\max_{\{\tilde{S}_i\}_{i=1}^K : \tilde{S}_i \succeq 0, \sum_{i=1}^K \mathrm{Tr}(\tilde{S}_i) \leq P} \sum_{i=1}^{K} \log |I + \Lambda_i^{\frac{1}{2}} \tilde{S}_i \Lambda_i^{\frac{1}{2}}|, \tag{3.31}$$

where there is a unitary matrix $U_i^{(\tilde{n})}$ such that

$$\left(U_i^{(\tilde{n})}\right)^\dagger G_i^{(\tilde{n})} \left(G_i^{(\tilde{n})}\right)^\dagger U_i^{(\tilde{n})} = \Lambda_i = \mathrm{diag}\left((\Lambda_i)_{1,1}, \cdots, (\Lambda_i)_{n,n}\right), \forall i. \tag{3.32}$$

Let $\left\{(\Lambda_i)_{1,1}, \cdots, (\Lambda_i)_{n,n}\right\}$ be ordered monotonically decreasing into $\{\lambda_i\}_{i=1}^{K}$. Further, utilizing the Hadamard Determinantal Inequality [7], i.e., if $A = (a_{ij})_{n \times n}$ is a real (or Hermitian) positive semidefinite matrix, then $\det(A) \leq a_{11} \cdots a_{nn}$; this equality holds if and only if A is diagonal, the optimization problem (3.31) is equivalent to the following problem:

$$\begin{aligned} \max_{\left\{(\tilde{S}_i)_{t,t}\right\}} & \sum_{i=1}^{K} \sum_{t=1}^{n} \log\left(1 + (\Lambda_i)_{t,t}\left(\tilde{S}_i\right)_{t,t}\right) \\ \text{subject to } & \sum_{i=1}^{K} \sum_{t=1}^{n} \left(\tilde{S}_i\right)_{t,t} \leq P; \left(\tilde{S}_i\right)_{s,t} = 0, \\ & \text{as } s \neq t, 1 \leq s, t \leq n; i = 1, \ldots, K, \end{aligned} \tag{3.33}$$

and $\overline{Q}_i^{(\tilde{n})} = U_i^{(\tilde{n})} \overline{S}_i^* \left(U_i^{(\tilde{n})}\right)^\dagger$, where $\left\{\overline{Q}_i^{(\tilde{n})}\right\}_{i=1}^{K}$ and $\left\{\overline{S}_i^*\right\}_{i=1}^{K}$ are the optimal solutions of the optimization problem (3.27) and (3.33), respectively.

According to the first step in the step 3 of IWFAwFIS for IGWF-BC, the set

$$\left\{(S_i^*)_{t,t}, 1 \leq t \leq k_i^*\right\}, \forall i,$$

implies

$$\frac{1}{\frac{1}{(\Lambda_i)_{t,t}} + (S_i^*)_{t,t}} = \frac{1}{\frac{1}{\lambda_{k^*}} + S_{k^*}}, 1 \leq t \leq k_i^*, \forall i.$$

There is a number μ which satisfies the following equalities.

$$\mu = \frac{1}{\frac{1}{(\Lambda_i)_{t,t}} + (S_i^*)_{t,t}} = \frac{1}{\frac{1}{\lambda_{k^*}} + S_{k^*}}.$$

Let the number $\sigma_{i,t} = 0, t = 1, \ldots, k_i^*, \forall i$. Due to, as $k_i^* < t \le n$,

$$\mu = \frac{1}{\frac{1}{\lambda_{k^*}} + S_{k^*}} \ge (\Lambda_i)_{t,t}, \; \sigma_{i,t} = \mu - (\Lambda_i)_{t,t} \ge 0.$$

the Lagrange function of the problem (3.33) is

$$L\left(\left\{(S_i)_{t,t}\right\}; \overline{\mu}, \{\overline{\sigma}_{i,t}\}\right) = \sum_{i=1}^{K} \sum_{t=1}^{n} \log\left(1 + (\Lambda_i)_{t,t} (S_i)_{t,t}\right)$$
$$+ \overline{\mu}\left(P - \sum_{i=1}^{K} \sum_{t=1}^{n} (S_i)_{t,t}\right) + \sum_{i=1}^{K} \sum_{t=1}^{n} \overline{\sigma}_{i,t} (S_i)_{t,t},$$

$$(3.34)$$

where $\overline{\mu}$ and $\{\overline{\sigma}_{i,t}\}$ are the Lagrange multipliers. It is easily known that the set $\{S_{t,t}^*\}$, μ and $\{\sigma_{i,t}\}$ mentioned above satisfy the KKT conditions of the equivalent model, i.e.,

$$\begin{cases} \frac{1}{\frac{1}{(\Lambda_i)_{t,t}} + (S_i^*)_{t,t}} - \mu + \sigma_{i,t} = 0, & 1 \le t \le n \\ (S_i^*)_{t,t} \ge 0, \; \sigma_{i,t} (S_i^*)_{t,t} = 0, \; \sigma_{i,t} \ge 0, \; \forall i, t & (3.35) \\ \sum_{i=1}^{K} \sum_{t=1}^{n} (S_i^*)_{t,t} = P, & \mu \in \mathbb{R}. \end{cases}$$

Due to the problem (3.33) being a convex optimization problem with a concave objective function, $\{S_i^*, i = 1, \ldots, K\}$ is an optimal solution to the problem (3.33). Due to the equivalence between the problem (3.27) and the problem (3.33), $\left\{Q_i^{(n)}\right\}_{i=1}^{K}$, obtained by the step 3 of IWFAwFIS for IGWF-BC, is indeed the optimal solution to the problem (3.27).

Second, we apply the fixed point theory set up by us with a series of derivations, the convergence for the algorithm IGWF-BC is then obtained. To reduce the cost of computation, (2) of IGWF-BC utilizes Fibonacci search. To improve the performance of the algorithm and reduce the cost of computation for the sum rate optimization problem mentioned above, the objective function in (3) of IGWF-BC may only be evaluated at the four points of $\beta = \frac{1}{K}, \frac{1}{K} + \frac{1}{3}\left(1 - \frac{1}{K}\right), \frac{1}{K} + \frac{2}{3}\left(1 - \frac{1}{K}\right)$ and 1 by parallel computation to find β^*. The following numerical experiments show the improvement of the algorithm.

3.5 Numerical Examples for RRM in MIMO System

We first present two examples for MIMO MAC communications, and then two examples for MIMO BC ones.

3.5.1   Examples for RRM in MIMO MAC Wireless Communications

We start to show some numerical examples to illustrate the simplification and effectiveness of our algorithms. For clear understanding, the iterative water-filling algorithm in [6] is called Algorithm A1 in this book.

Example 3.1. The performance of Algorithm IGWF-MAC, compared with Algorithm A1 of [6], is presented in Figs. 3.1 and 3.2, where $m = 8$ and $s = 8$. Random data are generated for the channel gain matrices. The number of the users are 8, 18, 28, 38, respectively. The individual power constraints are $P_j = 2, j = 1, 2, \ldots, K$.

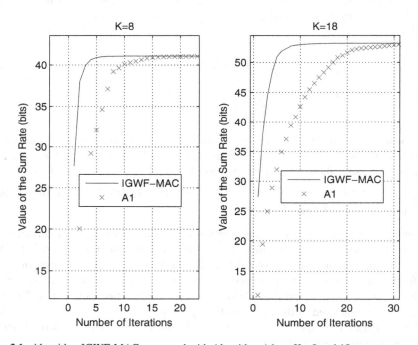

Fig. 3.1 Algorithm IGWF-MAC compared with Algorithm A1, as K = 8 and 18

From Figs. 3.1 and 3.2, the solid curves and the cross markers represent the results of our proposed algorithm and Algorithm A1 respectively. These results show that our proposed algorithm exhibits faster convergence than that of A1. Especially with the increasing of the number of users, the gain is more significant.

On one hand, the above figures are plotted by the same way of [6] when algorithms are compared; on the other hand, prior to **Algorithm IGWF-MAC** being proposed, Algorithm A1 [6] was the most rapidly convergent algorithm compared to others. This point has been emphasized, as an important characteristic in [6]. However, since IGWF-MAC can utilize parallel computation and choose the

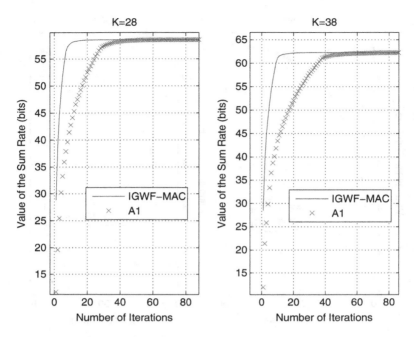

Fig. 3.2 Algorithm IGWF-MAC compared with Algorithm A1, as K = 28 and 38

Fibonacci search, not only it reduces the computation amount, but also it greatly speeds up the convergent rate. For an iteration for IGWF-MAC, it evaluates the objective function by parallel computation and also utilizes Fibonacci search twice. Although A1 does not need to evaluate the objective function during the inner loop, it carries out more iterations to reach the equal increment of the objective function. Thus, since every iteration during the inner loop for A1 has to implement the inverse operation of the matrix to compute the effective channel matrix, the eigenvalues and eigenvectors, for computing the unitary matrix $U_k^{(n)}$, A1 has to take on more computation burden than IGWF-MAC. Furthermore, as the number of the users increases, the advantages from IGWF-MAC appear more remarkably on the computation time and amount, and convergent rate over A1.

Therefore, the advantage of Algorithm IGWF-MAC is seen clearly with its theoretical advance.

3.5.2 Examples for RRM in MIMO BC Wireless Communications

We start to show two numerical examples to illustrate the simplification and effectiveness of our algorithm.

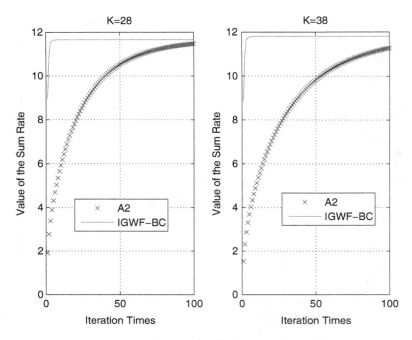

Fig. 3.3 Algorithm IGWF-BC compared with Algorithm A2, as K = 28 and 38

Example 3.2. The performance of Algorithm IGWF-BC, compared with Algorithm A2 of [5], is presented in Figs. 3.3 and 3.4, where $m = 8$ and $n = 8$. Random data are generated for the channel gain matrices. The number of the users are 28, 38, 58 and 158, respectively. The sum power constraint is $P = 2$.

From Figs. 3.3 and 3.4, the upper and lower curves represent the results of the proposed algorithm and that of A2 respectively. It is shown that Algorithm IGWF-BC outperforms Algorithm A2 in terms of convergence speed. Especially when the number of the users increases, the gain from Algorithm IGWF-BC is more significant.

A hybrid algorithm was presented in [5] and claimed to be outperforming any other alternatives. It is referred here as "Original + Algorithm 2". Our proposed algorithm exhibits improved performance as shown in the following example.

Example 3.3. The performance of Algorithm IGWF-BC, compared with the hybrid algorithm, i.e. Algorithm Original + Algorithm 2 (Original + A2) of [5], is presented in Figs. 3.5 and 3.6, where $m = 8$ and $n = 8$. Random data are generated for the channel gain matrices. The number of the users are 28, 38, 58 and 158, respectively. The sum power constraint is $P = 2$.

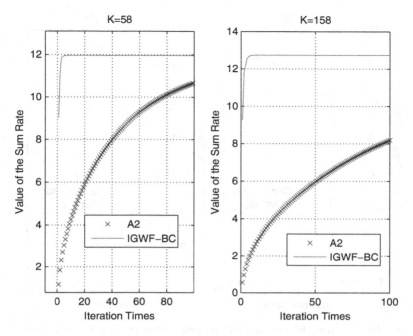

Fig. 3.4 Algorithm IGWF-BC compared with Algorithm A2, as K = 58 and 158

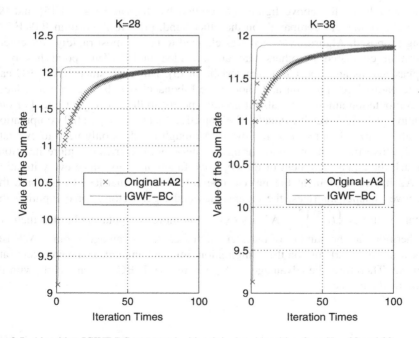

Fig. 3.5 Algorithm IGWF-BC compared with original + Algorithm 2, as K = 28 and 38

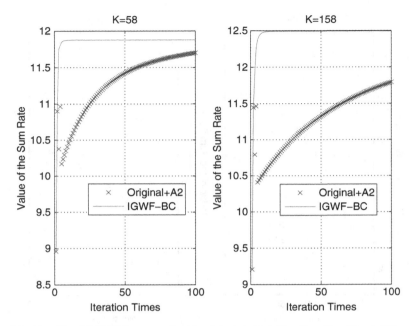

Fig. 3.6 Algorithm IGWF-BC compared with original + Algorithm 2, as K = 58 and 158

On one hand, the above figures are plotted by the same way of [5] and [9] when algorithms are compared; on the other hand, prior to Algorithm IGWF-BC being proposed, Algorithm A2 [5] was claimed to be the most rapidly convergent algorithm compared to others that are included in [5]. This point has been emphasized, as an important characteristic in [5]. However, since IGWF-BC can utilize parallel computation and choose the Fibonacci search, not only it reduces the computation amount, but also it greatly speeds up the convergent rate. For one iteration of IGWF-BC, it evaluates the objective function by parallel computation and also utilizes Fibonacci search twice. Although A2 does only need to evaluate the objective function one time during one iteration, it carries out more iterations to reach the equal increment of the objective function. Thus, since every iteration for A2 has to implement the inverse operations of the matrices to compute the effective channel matrices, the eigenvalues and eigenvectors, for computing the unitary matrices $\left\{ U_i^{(\bar{n})} \right\}_{i=1}^{K}$, A2 has to take on more computation burden than A1. Furthermore, as the number of the users increases, the advantages from IGWF-BC appear more remarkably on the computation time and amount, and convergent rate over A2. Therefore, the advantage of Algorithm IGWF-BC is seen clearly with its theoretical advance.

References

1. P. He and L. Zhao, "Improved sum power iterative water-filling with rapid convergence and robustness for multi-antenna Gaussian broadcast channels," IEEE VTC Spring, pp. 1–5, 2010.
2. P. He and L. Zhao, "Improved iterative water-filling with rapid convergence and parallel computation for Gaussian multiple access channels," IEEE VTC Spring, pp. 1–5, 2010.
3. E. Telatar, "Capacity of multi-antenna Gaussian channels," European Transactions on Telecommunications, vol. 10, pp. 585–596, 1999.
4. E. Biglieri, R. Calderbank, A. Constantinides, A. Goldsmith, A. Paulraj and H. V. Poor, MIMO Wireless Communications, Cambridge University Press, Cambridge, 2007.
5. N. Jindal, W. Rhee, S. Vishwanath, S. A. Jafar and A. Goldsmith, "Sum power iterative water-filling for multi-antenna Gaussian broadcast channels," IEEE Transactions on Information Theory, vol. 51, pp. 1570–1580, 2005.
6. W. Yu, W. Rhee, S. Boyd and J. M.Cioffi, "Iterative water-filling for Gaussian vector multi-access channels," IEEE Transactions on Information Theory, vol. 50, pp. 145–152, 2004.
7. J. R. Magnus and H Neudecker, Matrix Differential Calculus with Applications in Statistics and Econometrics, 2nd Edition, Wiley, 1999.
8. P. He and L. Zhao, "Correction of convergence proof for iterative water-filling in Gaussian MIMO broadcast channels," IEEE Transactions on Information Theory, vol. 57, pp. 2539–2543, 2011.
9. W. Yu, "Sum-capacity computation for the Gaussian vector broadcast channel via dual decomposition," IEEE Transactions on Information Theory, vol. 52, pp. 754–759, 2006.
10. W. Zangwill, Nonlinear Programming: A Unified Approach, Prentice-Hall, Englewood Cliffs, 1969.
11. M. Kobayashi and G. Caire, "An iterative water-filling algorithm for maximum weighted sum-rate of Gaussian MIMO-BC," IEEE Journal of Selected Areas in Communications, vol. 24, pp. 1640–1646, 2006.
12. P. Viswanath and D. Tse, "Sum capacity of the multiple antenna Gaussian broadcast channel and uplink-downlink duality," IEEE Transactions on Information Theory, vol. 49, pp. 1912–1921, 2003.
13. S. Vishwanath, N. Jindal and A. Goldsmith, "Duality, achievable rates and sum-rate capacity of Gaussian MIMO broadcast channels," IEEE Transactions on Information Theory, vol. 49, pp. 2658–2668, 2003.
14. N. Jindal, S. Vishwanath and A. Goldsmith, "On the duality of Gaussian multiple-access and broadcast channels," IEEE Transactions on Information Theory, vol. 50, pp. 768–783, 2004.

Chapter 4
RRM for Cognitive Network

The radio spectrum is a precious resource that demands efficient utilization as the currently licensed spectrum is severely underutilized [1]. Cognitive Radio (CR) [2–4], which adapts the radios operating characteristics to the real-time conditions, is the key technology that allows flexible, efficient and reliable spectrum utilization in wireless communications. This technology exploits the underutilized licensed spectrum of the primary user(s) (PU) and introduces the secondary user(s) (SU) to operate on the spectrum that is either opportunistically being available or concurrently being shared by the PU and the SU.

Under this situation and according to the definition of the cognitive (radio) network [5], opportunistically utilizing the spectrum is that the SUs may fill the spectrum gaps or holes left by the PUs; concurrently utilizing the spectrum is that the SUs transmit over the same spectrum as the PUs, in the way that the interference from the transmitting of the SUs not violating the transmitting quality requirement from the PUs. This chapter focuses on the latter case. Furthermore, the multiple-input multiple-output (MIMO) technology uses multiple antennas at either the transmitter or the receiver to significantly increase data throughput and link range without additional bandwidth or transmitted power. Thus it plays an important role in wireless communications today. Since for infrastructure-supported networks, such as the widely used cellular network, the base station(s) is often utilized for all the users, this usage is assumed in this chapter. In this chapter, we consider multiple SUs access the base station, referred as multiple access channel (MAC).

4.1 Introduction

In this chapter, we discuss the multi-user single input multiple output MAC system (SIMO-MAC) under CR network and its optimal power allocation solution to maximize the weighted sum-rate. The weighted sum-rate maximization problem is to compute the "best" achievable rate vector in the capacity region [6–8] by specifying the working point at the boundary of the capacity region. This optimality problem is of the Pareto meaning under multi-objective optimization.

For the non-CR cases, the sum-rate maximization problem has been intensively explored for both Gaussian broadcast channel (BC) [9, 10] and Gaussian MAC [11]. Typical approaches include iterative water-filling algorithm [9, 11] and dual-decomposition [10]. The conventional water-filling algorithm [12] which is an efficient resource allocation algorithm needs to be used inside each of the iterations as an inner loop operation. In addition, the set up of the well known duality between the Gaussian BC and the sum-power constrained Gaussian dual MAC [13–15] facilities the transform of BC sum-rate problems into its dual MAC problem. As for the weighted sum-rate problem, it is easily seen that as the weighted coefficients all being unity, the weighted sum-rate problem is reduced into a sum-rate optimization problem. Thus, solving the weighted sum-rate problem is more general. However, due to its more complicated structure, the conventional water-filling [12] might not be able to so compute its solution as it does in sum-rate calculation problem. For computing the maximum weighted sum-rate for a class of the Gaussian single-input multiple-output (SIMO) BC systems or equivalent dual MAC systems, Kobayashi and Caire [16] has presented some algorithms using a cyclic coordinate ascent algorithm to provide the max-stability policy.

For the CR cases, besides the individual power constraints to the SUs, the total interference power from the SUs needs to be included into the constraints of the target problem. Since single-antenna mobile users are quite common and compose a major served group due to the size and cost limitations of mobile terminals, this chapter is confined to a single input multiple output multiple access channel (SIMO-MAC) in the CR network. Earlier work [17, 18] investigated sum-rate problem and weighted sum-rate problem in CR-SIMO-MAC cases respectively. In addition, for the ergodic sum capacity of single input single out (SISO) system, Zhang et al. [19] studied the maximum (non-weighted) sum-rate problem with a simple form of the objective function.

In this chapter, by exploiting the structure of the weighted sum-rate optimization problem, we propose an efficient iterative algorithm to compute the optimal input policy and to maximize the weighted sum-rate, via solving a generalized water-filling in each of the iterations. The water-filling machinery is experiencing continuous development [12, 20–23]. We propose a geometric weighted water-filling algorithm (GWWFA) to form a fundamental step (inner loop algorithm) for the target problem. In the inner loop, the weighted sum-rate problem is decomposed into a series of generalized water-filling problems. With this decomposition, a decoupled system with each equation of the decoupled system containing only a scalar variable is formed and solved. Any one of the equations is solved by the GWWFA with a finite number of loops. To speed up the computation of the solution to each of the equations, we also specify the intervals the solution belongs to.

For the outer loop of the algorithm, variable scale factor is applied to update the covariance vector of the users. The optimal scale factor is obtained by maximizing the target objective value (i.e., the weighted sum-rate) in the scalar variable β to expedite convergence of the proposed algorithm. In order to achieve this purpose, we determine an optimal scale factor by searching in a range which consists of a few discrete values. As a result, parallel operation can be used together to expedite the search and to avoid the requirement of another nested loop. This parallel operation can be distributed to and carried out by the multiple processors (for example, four processors).

Compared with earlier work [18], the main difference of our work is that: (i) in [18], the dual-decomposition approach [10] is used. In our work, we apply the iterative water-filling algorithm [9] and extend the algorithm to solve the target problem. The advantage of the iterative water-filling algorithm is that it is a monotonic feasible operator to the iteration. That is to say, the proposed algorithm generates a sequence composed of feasible points in its iterations. The objective function values, corresponding to this point sequence, are monotonically increasing. Hence, the stop criterion for computation might be easily set up. However, the regular primal-dual method used in [18] is not a feasible point method; (ii) for the constraints of the target problem, we make the individual power constraints more strict and more reasonable, due to the values of signal powers being assumed to be greater than or equal to zero; (iii) the convergence rate is improved significantly. In the numerical example in [18] (Fig. 4.4), the convergence of the weighted sum-rate is obtained after 110 iterations for a system with 3 SUs and 2 PUs. However, with our proposed algorithm, we achieve the weighted sum-rate convergence with two iterations with the simulated range (number of SUs up to 110). In addition, even if the PUs and SUs are served by different base stations, it is easy to see that the proposed machinery can be used with some minor modifications.

In the remaining of this chapter, system model for a CR-SIMO-MAC system and its weighted sum-rate are described in Sect. 4.2. Section 4.3 discusses the proposed algorithm to solve the maximal weighted sum-rate problem through an inner loop algorithm and its convergence proof, presented in Sect. 4.3.1. Then the outer loop algorithm: iterative geometric water-filling for weighted CR cases (IGWF-WCR) and its implementation are presented in Sect. 4.3.2. Section 4.4 provides the convergence proof of the IGWF-WCR. Section 4.5 presents numerical results and some complexity analysis to show the effectiveness of the proposed algorithm.

Key notations that are used in this chapter are revisited as follows: $|\mathbf{A}|$ and $\mathrm{Tr}\,(\mathbf{A})$ give the determinant and the trace of a square matrix $\mathbf{A}$, respectively; $E[X]$ is the expectation of the random variable X; the capital symbol $\mathbf{I}$ for a matrix denotes the identity matrix with the corresponding size. A square matrix $\mathbf{B} \succeq 0$ means that $\mathbf{B}$ is a positive semi-definite matrix. Further, for arbitrary two positive semi-definite matrices $\mathbf{B}$ and $\mathbf{C}$, the expression $\mathbf{B} \succeq \mathbf{C}$ means that the difference: $\mathbf{B} - \mathbf{C}$ is a positive semi-definite matrix. In addition, for any complex matrix, its superscripts $\dagger$ and T denote the conjugate transpose and the transpose of the matrix, respectively.

4.2 SIMO-MAC in Cognitive Radio Network and Its Weighted Sum-Rate

For a SIMO-MAC in the CR network, as shown in Fig. 4.1, assume that there are one base station (BS) with N_r antennas, and K SUs and N PUs, each of which is equipped with one single antenna. The received signal $\mathbf{y} \in \mathbb{C}^{N_r \times 1}$ at the BS is described as

$$\mathbf{y} = \sum_{i=1}^{K} \mathbf{h}_i^\dagger \mathbf{x}^i + \sum_{i=1}^{N} \hat{\mathbf{h}}_i^\dagger \hat{\mathbf{x}}^i + \mathbf{Z}, \quad \mathbf{h}_i \in \mathbb{C}^{1 \times N_r}, i = 1, 2, \cdots, K, \text{ and}$$
$$\hat{\mathbf{h}}_i \in \mathbb{C}^{1 \times N_r}, i = 1, 2, \cdots, N, \tag{4.1}$$

where the i-th entry $\mathbf{x}^i$ of $\mathbf{x} \in \mathbb{C}^{K \times 1}$ is a scalar complex input signal from the i-th SU and $\mathbf{x}$ is assumed to be a Gaussian random vector having zero mean with independent entries. The j-th entry $\hat{\mathbf{x}}^j$ of $\hat{\mathbf{x}}$ is a scalar complex input signal from the j-th PU and $\hat{\mathbf{x}}$ is assumed to be a Gaussian random vector having zero mean with independent entries. The noise term, $\mathbf{Z} \in \mathbb{C}^{N_r \times 1}$ is an additive Gaussian noise random vector, i.e., $\mathbf{Z} \sim \mathbb{N}(0, \sigma^2 \mathbf{I})$. The channel input, $\hat{\mathbf{x}}, \mathbf{x}$ and $\mathbf{Z}$ are also assumed to be independent. Furthermore, the i-th SU's transmitted power can be expressed as

$$S_i \triangleq E\left[|\mathbf{x}^i|^2\right], i = 1, 2, \cdots, K. \tag{4.2}$$

Note that $S_i, \forall i$, is non-negative.

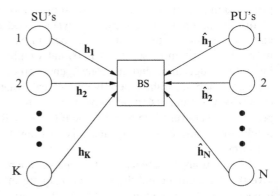

Fig. 4.1 CR-MAC system model

The mathematical model of the weighted sum-rate optimization problem for the SIMO-MAC in the CR network can be stated as follows (refer to (2.16) in [8] and therein):

Given a group of weights $\{w_k\}_{k=1}^K$ which is assumed to be in decreasing order (users can be arbitrarily renumbered to satisfy this condition) with the achievable rate of the secondary user k, $\log \frac{|\mathbf{C}_0 + \sum_{j=1}^{k} \mathbf{h}_j^\dagger \mathbf{h}_j S_j|}{|\mathbf{C}_0 + \sum_{j=1}^{k-1} \mathbf{h}_j^\dagger \mathbf{h}_j S_j|}, \forall k$, the weighted sum-rate is organized by

$$
\begin{aligned}
f_{wmac}\left(\mathbf{h}_1^\dagger, \cdots, \mathbf{h}_K^\dagger; P_1, \cdots, P_K; P_t\right) = \ & \max_{\{S_k\}_{k=1}^K} w_K \log\left|\mathbf{C}_0 + \sum_{j=1}^{K} \mathbf{h}_j^\dagger \mathbf{h}_j S_j\right| \\
& + \sum_{k=1}^{K-1}(w_k - w_{k+1}) \log\left|\mathbf{C}_0 + \sum_{j=1}^{k} \mathbf{h}_j^\dagger \mathbf{h}_j S_j\right| \\
& \text{Subject to: } 0 \le S_k \le P_k, \forall k; \sum_{k=1}^{K} g_k S_k \le P_t,
\end{aligned}
\tag{4.3}
$$

where, for the MAC cases, the peak power constraint on the kth SU exists and is denoted by a group of positive numbers: $P_k, k = 1, \ldots, K$; the power threshold to ensure the QoS of the PUs is denoted by the positive number P_t. Further, under no confusion, f_{dmac} is simply written as f. For convenience, we define η_k by $w_k - w_{k+1}$ for $k = 1, \ldots, K-1$; and η_K by w_K, as a group of non-negative real numbers, and assume one of them at least to be non-zero. Further, the term $g_k = \mathbf{h}_k \mathbf{h}_k^\dagger, \forall k$, is the channel gain of the kth SU to the BS. Also, we denote the covariance matrix of the random vector $\sum_{i=1}^{N} \hat{\mathbf{h}}_i^\dagger \hat{\mathbf{x}}^i + Z$ by $\mathbf{C}_0$, which is positive definite.

The constraint $\sum_{k=1}^{K} g_k S_k \le P_t$ is called the sum-power constraint with gains. The constraint is obtained in the following analysis. Let

$$
\mathbf{H} = \left[\mathbf{h}_1^\dagger, \cdots, \mathbf{h}_K^\dagger\right] \in \mathbb{C}^{N_r \times K} \text{ and } \hat{\mathbf{H}} = \left[\hat{\mathbf{h}}_1^\dagger, \cdots, \hat{\mathbf{h}}_N^\dagger\right] \in \mathbb{C}^{N_r \times N}.
\tag{4.4}
$$

Thus, the received signal at BS is $\mathbf{y} = \hat{\mathbf{H}}\hat{\mathbf{x}} + (\mathbf{H}\mathbf{x} + \mathbf{Z})$, where $\mathbf{H}\mathbf{x} + \mathbf{Z}$ can be regarded as the additive interference and noise to the transmitted signal $\hat{\mathbf{H}}\hat{\mathbf{x}}$ from the PUs. To guarantee the QoS for the PUs, the power of the interference and noise should be less than a threshold value, P_{TH}. This condition can be expressed as

$$
\text{Tr}(\mathbf{H}E(\mathbf{x}\mathbf{x}^\dagger)\mathbf{H}^\dagger + E(\mathbf{Z}\mathbf{Z}^\dagger)) \le P_{TH}.
\tag{4.5}
$$

It can be written as

$$
\sum_{k=1}^{K} g_k S_k \le P_{TH} - N_r \sigma^2 = P_t,
\tag{4.6}
$$

where the power constraint value P_t is the interference and noise threshold subtracted by the power of Gaussian noise.

As an alternative, to guarantee the QoS for each of the PUs, individually, the power of the interference and noise should be less than a threshold value, $P_{TH}(i), \forall i$. Similarly, it is obtained that

$$
\sum_{k=1}^{K} g_k S_k \le P_t(i), \forall i.
\tag{4.7}
$$

That is to say, the condition above is equivalent to

$$\sum_{k=1}^{K} g_k S_k \leq \min_i \{P_t(i)\}. \tag{4.8}$$

Name $\min_i\{P_t(i)\}$ as P_t and then the target model can still cover the case that the QoS for each of the PUs is considered. Note that at the base station with multiple antennas, the received signals can be regarded as a stochastic vector or point in a Hilbert space and the received signal powers are abstracted into the norm squared of the vector. The transmitted powers of the PUs have been taken into account by forming C_0 and P_t mentioned above, which appear in (4.3).

It is seen that $\{\eta_k\}$ stems from the vector of weights used in the multi-user information theory [8]. The parameter $\{\eta_k\}$ is called the weighted coefficients in the sequel without confusion.

A more strict weighted sum-rate model can also be obtained that reflects the essence of the issue for the CR-SIMO-MAC. Along a similar way mentioned above, we may choose the power thresholds $P_{t,i}$ to limit the impact from the SUs on each of the antennas of the BS. Thus the sum-power constraint with the gains is evolved into $\sum_{k=1}^{K} g_{k,i} S_k \leq P_{t,i}, i = 1, 2, \ldots, N_r$. It is seen that such an weighted sum-rate problem with more power constraints can be solved by solving a similar problem in (4.3). Therefore, this chapter aims at computing the solution to the problem (4.3). Note that if $\exists \mathbf{h}_{i_0} = \mathbf{0}, 1 \leq i_0 \leq K$, for (4.3), we remove the user i_0 and then the number of the users is reduced to $K - 1$. Thus, on this way, we can assume that $\mathbf{h}_i \neq \mathbf{0}, \forall i$.

For the sum-rate problem, as the important subset the problem (4.3) includes, assume that $M = \text{rank}(\mathbf{H})$. Applying the QR decomposition, $\mathbf{H} = \mathbf{QR}$. Let $\mathbf{Q} = [\mathbf{q}_1, \cdots, \mathbf{q}_M] \in \mathbb{C}^{N_r \times M}$ have orthogonal and normalized column vectors. $\mathbf{R} \in \mathbb{C}^{M \times K}$ is an upper triangle matrix with $r_{i,j}$ denoting the (i,j)-th entry of the matrix $\mathbf{R}$. $\mathbf{Q}^{\dagger}$ is regarded as an equalizer to the received signal by the BS. Thus, the i-th SU should have the rate:

$$R_i^{rate} = \log\left(1 + \frac{|r_{i,i}|^2 S_i}{\sigma^2 + \sum_{n=1}^{N} \widehat{S}_n \mathbf{q}_i^{\dagger} \hat{\mathbf{R}}_n \mathbf{q}_i + \sum_{j=i+1}^{K} |r_{i,j}|^2 S_j}\right), \tag{4.9}$$

where $\widehat{S}_n = E\left[\hat{\mathbf{x}}^n(\hat{\mathbf{x}}^n)^{\dagger}\right]$ and $\hat{\mathbf{R}}_n = \hat{\mathbf{h}}_n^{\dagger} \hat{\mathbf{h}}_n, n = 1, \cdots, N$. It is easy to see that the rate just mentioned comes from the expression: $\log \frac{|\mathbf{I} + \sum_{j=1}^{k} \mathbf{h}_j^{\dagger} \mathbf{h}_j S_j|}{|\mathbf{I} + \sum_{j=1}^{k-1} \mathbf{h}_j^{\dagger} \mathbf{h}_j S_j|}, \forall k$.

4.3   Algorithm IGWF-WCR

The proposed algorithm for the weighted sum-rate problem in the cognitive radio network, denoted by IGWF-WCR, uses variable scale factors as innovation and the generalized G in each iteration. In this section, firstly, the generalized geometric weighted water-filling problem and the proposed GWWFA are introduced, and the latter will be used as the inner loop algorithm of the IGWF-WCR. Then, the proposed IGWF-WCR and its implementation are presented.

4.3.1   Preparation of IGWF-WCR

Being a fundamental block of the optimum resource allocation problem for the CR-SIMO-MAC systems, the generalized water-filling problem is abstracted as follows.

For a multiple receiving antenna system, it is given that $P_t > 0$, as the total power or volume of the water; the allocated power and the propagation path (non-negative) gains for the ith user are given as S_i for $i = 1, \ldots, K$, and $\{a_{ij}\}_{j=i}^{K}$ respectively; and K is the total number of the users, to find that

$$\max_{\{S_i\}_{i=1}^{K}:\, 0 \leq S_i \leq P_i,\, \forall i;\, \sum_{i=1}^{K} g_i S_i \leq P_t} \sum_{i=1}^{K} \eta_i \sum_{j \in \{1,\ldots,i\}} \log(1 + a_{ij} S_j), \qquad (4.10)$$

where the set $\{\eta_i\}_{i=1}^{K}$ plays the role of the weighted coefficients. Note, as $\sum_{i=1}^{K} g_i P_i \leq P_t$, the solution to problem (4.3) is regressed into a trivial case. Hence, $\sum_{i=1}^{K} g_i P_i > P_t$ is assumed.

It is easy to see that if $a_{ij} = 0$, as $i \neq j$, and $P_i \gg 0, \forall i$, then the problem (4.10) is reduced into the conventional weighted water-filling problem, where the expression: $P_i \gg 0$ means that P_i is large enough. Further, if equal weights are chosen, it is reduced into conventional water-filling problem, which can be solved by the conventional water-filling algorithm [12].

To find the solution to the more complicated generalized problem above, the generalized geometric weighted water-filling algorithm (GWWFA) is presented as follows. Let

$$\lambda_i = \frac{1}{g_i} \sum_{j=i}^{K} \eta_j a_{ji}, \quad i = 1, \cdots, K. \qquad (4.11)$$

Utilize a permutation operation π on $\{\lambda_i\}$ such that

$$\lambda_{\pi(1)} \geq \lambda_{\pi(2)} \geq \cdots \geq \lambda_{\pi(K)} > \min_{1 \leq i \leq K} \left\{ v \,\middle|\, v = \frac{1}{g_i} \sum_{j=i}^{K} \frac{\eta_j a_{ji}}{1 + a_{ji} P} > 0 \right\} = \lambda_{\pi(K+1)}, \qquad (4.12)$$

where $P = \sum_{k=1}^{K} P_k$. Define function $J_i(s_i)$ as

$$J_i(s_i) = \frac{1}{g_i} \sum_{j=i}^{K} \frac{\eta_j a_{ji}}{1 + a_{ji} s_i}, \quad i = 1, \cdots, K. \qquad (4.13)$$

It is easy to see that the function $J_i(s_i)$ is strictly monotonically decreasing and continuous over the interval

$$\left(-\min_j\left\{\frac{1}{a_{ji}}\middle|a_{ji}>0\right\},\infty\right). \tag{4.14}$$

The steps of the GWWFA can be described as below.

Algorithm GWWFA:

1. Given $\varepsilon > 0$, initialize $\lambda_{\min}$ and $\lambda_{\max}$.
2. Set $\lambda = (\lambda_{\min} + \lambda_{\max})/2$.
3. If λ falls in the interval $[\lambda_{\pi(i+1)}, \lambda_{\pi(i)}]$, where $1 \leq i \leq K$, initialize the point $\left[s_{\pi(1)}^{(0)}, \cdots, s_{\pi(i)}^{(0)}\right]$ and compute

$$\left[s_{\pi(1)}^{(n+1)}, \cdots, s_{\pi(i)}^{(n+1)}\right] = \left[s_{\pi(1)}^{(n)} - \frac{J_{\pi(1)}\left(s_{\pi(1)}^n\right) - \lambda}{J_{\pi(1)}'\left(s_{\pi(1)}^n\right)}, \cdots, s_{\pi(i)}^{(n)} - \frac{J_{\pi(i)}\left(s_{\pi(i)}^n\right) - \lambda}{J_{\pi(i)}'\left(s_{\pi(i)}^n\right)}\right]. \tag{4.15}$$

Then $n <= n+1$. Repeat the procedure in (4.15) until the point $\left[s_{\pi(1)}^{(n)}, \cdots, s_{\pi(i)}^{(n)}\right]$ converges. Denote $\lim_n \left[s_{\pi(1)}^{(n)}, \cdots, s_{\pi(i)}^{(n)}\right]$ by $\left[s_{\pi(1)}^*, \cdots, s_{\pi(i)}^*\right]$. Let

$$\left[s_{\pi(i+1)}^*, \cdots, s_{\pi(K)}^*\right] = 0 \in \mathbb{R}^{1 \times (K-i)}.$$

4. If $\sum_{k=1}^K g_k s_k^* - P_t > 0$, then $\lambda_{\min}$ is assigned λ;
   if $\sum_{k=1}^K g_k s_k^* - P_t < 0$, then $\lambda_{\max}$ is assigned λ;
   If $\sum_{k=1}^K g_k s_k^* - P_t = 0$, stop.
5. If $|\lambda_{\min} - \lambda_{\max}| \leq \varepsilon$, stop. Otherwise, goto step 2.

Remark 4.1. Note, in (1) of the GWWFA, that the initial $\lambda_{\min}$ may be chosen as $\lambda_{\pi(K+1)}$, and $\lambda_{\max}$ may be chosen as $\lambda_{\pi(1)}$.

In (3), for the initialization of $s_{\pi(k)}^{(0)}$, first, we may choose an interval, such as $[0, P_{\pi(k)}]$, and use the secant method or the bisection method [24] over the interval to compute, in parallel, an approximate solution to the system $J_{\pi(k)}(s_{\pi(k)}) - \lambda = 0, \forall k$. Hence, only through a few loops ($\leq \lceil \log_2 P_{\pi(k)} \rceil + 1$ loops), $|e_0|$, as an absolute error between the accurate solution and the approximate solution obtained by the method above, is less than 0.5. The initialization of $s_{\pi(k)}^{(0)}$, for $k = 1, \ldots, i$, is assigned by the above approximate solution. Let $(e_n)_k = s_{\pi(k)}^* - s_{\pi(k)}^{(n)}$, where $J(s_{\pi(k)}^*) - \lambda = 0$. It is seen that

$$(e_{n+1})_k = (e_n)_k^2 \frac{\sum_{j=\pi(k)}^{K} \frac{1}{8_{\pi(k)}} \eta_j a_{j\pi(k)}^2 \frac{1}{(1+a_{j\pi(k)}(s_{\pi(k)}^* - (e_n)_k))^2} \frac{1}{1+a_{j\pi(k)} s_{\pi(k)}^*}}{\sum_{j=\pi(k)}^{K} \frac{1}{8_{\pi(k)}} \eta_j a_{j\pi(k)}^2 \frac{1}{(1+a_{j\pi(k)}(s_{\pi(k)}^* - (e_n)_k))^2}} = (e_n)_k^2 \rho_n, \qquad (4.16)$$

where $0 < \rho_n < 1$. It can be observed that $0 \le (e_m)_k < (e_0)_k^{2^m}$ and then $\{s_{\pi(k)}^{(m)}\}_{m=1}^{\infty}$ uniformly converges, $0 \le (e_6)_k < 10^{-19}$ (machine zero), $\forall k$. That is to say, the absolute error between the approximative solution and the accurate solution is the machine zero within six loops. Thus, the optimal solution $(s_{\pi(1)}^*, \cdots, s_{\pi(i)}^*)$ can be obtained in parallel, within finite loops.

Denote a function

$$(x)_0^a = \begin{cases} 0, x < 0 \\ x, 0 \le x \le a \\ a, x > a. \end{cases} \qquad (4.17)$$

Then define $G(\lambda)$ as

$$G(\lambda) = \sum_{k=1}^{K} g_{\pi(k)} \left(J_{\pi(k)}^{-1}(\lambda) \right)_0^{P_{\pi(k)}}. \qquad (4.18)$$

Since $J_{\pi(k)}(s_{\pi(k)})$ is strictly monotonically decreasing and continuous over the interval, so are $J_{\pi(k)}^{-1}(\lambda)$ and $G(\lambda)$ over the corresponding interval(s). Due to $G(\lambda_{\pi(K+1)}) > P_t$ and $G(\lambda_{\pi(1)}) < P_t$, step 4 can make λ converge such that $G(\lambda) = P_t$. Optimality of the GWWFA is stated by following proposition.

Proposition 4.1. *For (4.10), its optimal solution can be obtained by the GWWFA.*

Proof of Proposition 4.1. From the third item of (4) and (5), $G(\lambda) = P_t$. Then

$$\sum_{k=1}^{K} g_{\pi(k)} \left(J_{\pi(k)}^{-1}(\lambda) \right)_0^{P(\pi(k))} = P_t. \qquad (4.19)$$

Since there exists $i_0: 1 \le i_0 \le K$ such that $\lambda \in [\lambda_{\pi(i_0+1)}, \lambda_{\pi(i_0)}], \underline{\mu}_{\pi(j)} = 0$ and $\overline{\mu}_{\pi(j)} = \lambda_{\pi(j)} - \lambda \ge 0$, as $j = 1, \ldots, i_0$; $\underline{\mu}_{\pi(j)} = \lambda - \lambda_{\pi(j)} \ge 0$ and $\overline{\mu}_{\pi(j)} = 0$, as $j = i_0 + 1, \ldots, K$. Therefore, there exists the solution

$$\left\{ s_{\pi(k)}^* (= (J_{\pi(k)}^{-1}(\lambda))_0^{P(\pi(k))}) \right\}_{k=1}^{K}, \qquad (4.20)$$

and the Lagrange multipliers λ, $\{\underline{\mu}_{\pi(k)}\}$ and $\{\overline{\mu}_{\pi(k)}\}$ mentioned above such that the KKT condition of the problem (4.10) holds, where the λ corresponds to the constraint $\sum_{k=1}^{K} g_k s_k \leq P_t$, $\{\underline{\mu}_{\pi(k)}\}$ and $\{\overline{\mu}_{\pi(k)}\}$ correspond to the constraints $\{s_{\pi(k)} \geq 0\}$ and $\{s_{\pi(k)} \leq P_{\pi(k)}\}$, respectively.

Since the problem in Proposition 4.1 is a differentiable convex optimization problem with linear constraints, not only is the KKT condition mentioned above sufficient, but it is also necessary for optimality. Note that it is easily seen that the constraint qualification (the CQ) of the optimization problem (4.10) holds. Proposition 4.1 hence is proved.

Remark 4.2. To decouple the variables in the objective function of the problem (4.3), a sum expression is acquired by adding the objective function, just mentioned, K times. Then the sum expression is operated, by one variable being selected as an optimized variable with respect to the others being fixed. Thus, from the expression (4.3), the problem (4.21),

$$\max_{\{S_l\}_{l=1}^{K}:0\leq S_l\leq P_l,\ \sum_{i=1}^{K} g_i S_i \leq P_t} \sum_{i=1}^{K} \eta_i \sum_{l=1}^{i} \log\left(1 + \mathbf{G}_{il}\left(\mathbf{G}_{il}\right)^{\dagger} S_l\right), \quad (4.21)$$

is implied as follows:

Since

$$\sum_{j=1}^{K}\sum_{i=1}^{K} \eta_i \log\left|\mathbf{C}_0 + \sum_{l\in\{1,\cdots,i\}\cap\{j\}} \mathbf{h}_l^{\dagger}\mathbf{h}_l S_l + \sum_{k\in\{1,\cdots,i\}\backslash\{j\}} \mathbf{h}_k^{\dagger}\mathbf{h}_k \overline{S}_k\right|$$

$$= \sum_{i=1}^{K} \eta_i \sum_{j=1}^{K} \log\left(1 + \sum_{l\in\{1,\cdots,i\}\cap\{j\}} \mathbf{G}_{il}\mathbf{G}_{il}^{\dagger} S_l\right) \quad (4.22)$$

$$+ \sum_{i=1}^{K} \eta_i \sum_{j=1}^{K} \left|\mathbf{C}_0 + \sum_{k\in\{1,\cdots,i\}\backslash\{j\}} \mathbf{h}_k^{\dagger}\mathbf{h}_k \overline{S}_k\right|, \quad (4.23)$$

where $\overline{S}_k, \forall k$, is fixed and

$$\mathbf{G}_{il} = \mathbf{h}_i \left(\mathbf{C}_0 + \sum_{k\in\{1,\cdots,i\}\backslash\{l\}} \mathbf{h}_k^{\dagger}\mathbf{h}_k \overline{S}_l\right)^{-\frac{1}{2}}, \forall i, l, \quad (4.24)$$

the optimization problem

$$\max_{\{S_k\}_{k=1}^{K}:\ 0\leq S_k\leq P_k,\ \sum_{k=1}^{K} g_k S_k \leq P_t} \sum_{j=1}^{K}\sum_{i=1}^{K} \eta_i \log\left|\mathbf{C}_0 + \sum_{l\in\{1,\cdots,i\}\cap\{j\}} \mathbf{h}_l^{\dagger}\mathbf{h}_l S_l\right.$$

$$\left. + \sum_{k\in\{1,\cdots,i\}\backslash\{j\}} \mathbf{h}_k^{\dagger}\mathbf{h}_k \overline{S}_k\right| \quad (4.25)$$

is equivalent to the problem below:

$$\max_{\{S_l\}_{l=1}^{K}: 0 \leq S_l \leq P_l, \; \sum_{i=1}^{K} g_i S_i \leq P_t} \sum_{i=1}^{K} \eta_i \sum_{l=1}^{i} \log \left(1 + \mathbf{G}_{il} \left(\mathbf{G}_{il} \right)^{\dagger} S_l \right). \tag{4.26}$$

If the CR SIMO weighted case is changed to the CR MIMO weighted case, it is a question whether there exists the fast water-filling like the algorithm mentioned above.

4.3.2 Algorithm IGWF-WCR and Its Implementation

The proposed algorithm IGWF-WCR, which is based on the combined problem of both the MIMO MAC and the CR network, is listed below.

Algorithm IGWF-WCR:

Input: vector $\mathbf{h}_i$, $S_i^{(0)} = 0$, $i = 1, \ldots, K; n = 1$.

1. Generate effective channels $\mathbf{G}_{ij}^{(n)} = \mathbf{h}_i \left(\mathbf{I} + \sum_{l \in \{1, \cdots, i\} \setminus \{j\}} \mathbf{h}_l^{\dagger} \mathbf{h}_l S_j^{(n-1)} \right)^{-\frac{1}{2}}$, for $i = 1, \ldots, K$, where the superscript with a pair of bracket, (n), represents the number of iterations.
2. Treating these effective channels as parallel, noninterfering channels, the new covariances $\left\{ \tilde{S}_i^{(n)} \right\}_{i=1}^{K}$ are generated by the GWWFA under the sum power constraint P_t. That is to say, $\left\{ \tilde{S}_i^{(n)} \right\}_{i=1}^{K}$ is the optimal solution to (4.27):

$$\max_{\{S_i\}_{i=1}^{K}: 0 \leq S_i \leq P_i, \; \sum_{i=1}^{K} g_i S_i \leq P_t} \sum_{i=1}^{K} \eta_i \sum_{j=1}^{i} \log \left(1 + \mathbf{G}_{ij}^{(n)} \left(\mathbf{G}_{ij}^{(n)} \right)^{\dagger} S_j \right). \tag{4.27}$$

Note that (4.27) is similar to (4.21), only $S_i^{(n-1)}$ and $\mathbf{G}_{ij}^{(n)}$ in the former take place of $\overline{S}_i$ and $\mathbf{G}_{il}$ in the latter, respectively, for any i, j, l.

3. Update step: Let $\gamma^{(n)}$ and $p^{(n-1)}$ denote the newly obtained covariance set and the immediate past covariance set respectively,

$$\gamma^{(n)} \triangleq \left(\tilde{S}_1^{(n)}, \tilde{S}_2^{(n)}, \cdots, \tilde{S}_K^{(n)} \right) \text{ and } p^{(n-1)} \triangleq \left(S_1^{(n-1)}, S_2^{(n-1)}, \cdots, S_K^{(n-1)} \right). \tag{4.28}$$

Let

$$\beta^* = \max \left\{ \beta_1 \Big| \beta_1 \in \arg \max_{\beta \in [1/K, 1]} f \left(\beta \gamma^{(n)} + (1 - \beta) p^{(n-1)} \right) \right\}, \tag{4.29}$$

as the innovation, where the function f has been defined in (4.3). Then, the covariance update step is

$$p^{(n)} = \left(S_1^{(n)}, S_2^{(n)}, \cdots, S_K^{(n)} \right) = \beta^* \gamma^{(n)} + (1 - \beta^*) p^{(n-1)}. \qquad (4.30)$$

The updated covariance is a convex combination of the newly obtained covariance and the immediate past covariance.

4. Let $n \mathrel{<=} n + 1$. Note it has been mentioned that $<=$ is the assignment operator. Go to **1** until convergence.

Note that the new algorithm employs variable weighting factors, which are obtained to maximize the objective function and to update the covariance.

In this section, the optimality of $\left\{ \tilde{S}_i^{(n)} \right\}_{i=1}^{K}$ has been proved, i.e., $\left\{ \tilde{S}_i^{(n)} \right\}_{i=1}^{K}$ is the solution to (4.21), by Proposition 4.1.

Remark 4.3. Due to the objective function $f\left(\beta \gamma^{(n)} + (1 - \beta) p^{(n-1)} \right)$ in Step 3 of Algorithm IGWF-WCR being (upper) convex, i.e., being concave, in the scalar variable β, for computing the maximum solution to the corresponding optimization problem, we can choose finite searching steps with even fewer evaluations of the objective function. Without loss of generality, the objective function in step (3) is evaluated at the four points $\left\{ \beta = \frac{1}{K}, \frac{1}{K} + \frac{1}{3} \left(1 - \frac{1}{K} \right), \frac{1}{K} + \frac{2}{3} \left(1 - \frac{1}{K} \right) \text{ and } 1 \right\}$ by parallel computation to determine β^*. That is to say, this parallel operation can be distributed to and carried out by the multiple processors (for example, four processors) of the base station, in order to expedite convergence of the proposed algorithm. Finally, the obtained satisfied solution is then distributed or returned to the corresponding secondary users.

The following Figs. 4.2–4.5 illustrate the proposed algorithms and their descriptions are detailed in Sect. 4.5.

4.4 Optimality of the Proposed Algorithm

In this chapter, as a more general model, we eliminate the assumption in [9] that the optimal solution is unique to prove convergence of the proposed algorithm. To the best knowledge of the authors, this is one of the proposed novelty for convergence of this class of algorithm with the spacer step [25] (page 125). Since our convergence proof is based on more general functions including an objective function and a few constraint functions, it will also enrich the optimization theory and methods. It is assumed that a mapping projects a point to a set. First, two concepts are introduced. The first concept is of an image of a mapping (or algorithm) that projects a point to a set; the second one is of a fixed point under the mapping (algorithm). Then, two lemmas are proposed, followed by the convergence proof of the proposed algorithm.

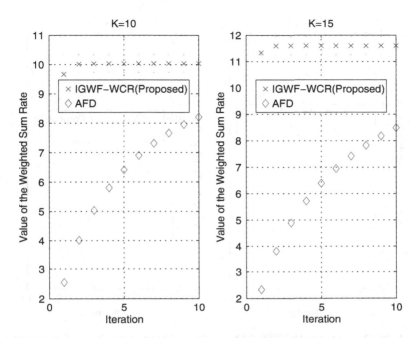

Fig. 4.2 Weighted sum-rate by IGWF-WCR and that by AFD with unit: bits, as K = 10 and 15

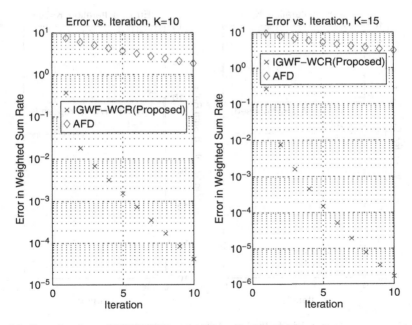

Fig. 4.3 Error functions of IGWF-WCR and AFD, as K = 10 and 15

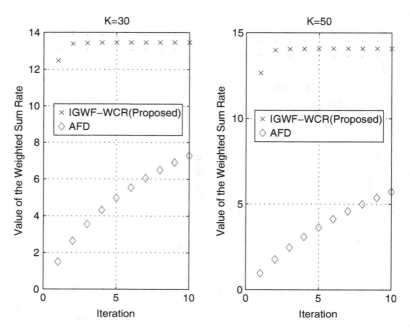

Fig. 4.4 Weighted sum-rate by IGWF-WCR and that by AFD with unit: bits, as K = 30 and 50

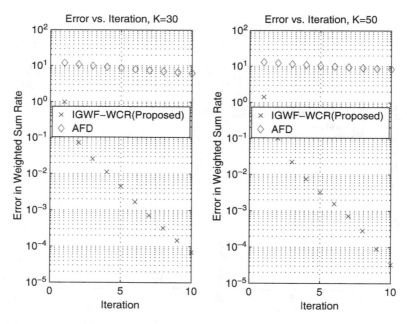

Fig. 4.5 Error functions of IGWF-WCR and AFD, as K = 30 and 50

Definition 4.1 (Image under Mapping or Algorithm A). (see e.g. [25], page 84)
Assume that X and Y are two sets. Let A be a mapping or an algorithm from X to Y,
which projects from a point in X to a set of points in Y. If the point in X is denoted
by x and the set of the points in Y is denoted by $A(x)$, then $A(x)$ is called the image
of x under A.

Definition 4.2 (Fixed Point under Mapping or Algorithm A). Let A be a mapping or an algorithm from X to Y. Assume $x \in X$. If $x \in A(x)$, x is said to be a fixed
point under A.

Note that (4.21) can be changed into a general form:

$$\left\{ \tilde{S}_i^{(n)} \right\}_{i=1}^{K} \in \arg \max_{\{S_i\}_{i=1}^{K}: 0 \leq S_i \leq P_i, \ \sum_{i=1}^{K} g_i S_i \leq P_t} \sum_{i=1}^{K} \eta_i \sum_{j=1}^{i} \log \left(1 + \mathbf{G}_{ij}^{(n)} \left(\mathbf{G}_{ij}^{(n)} \right)^{\dagger} S_j \right),$$

$$(4.31)$$

due to the condition of the optimal solution uniqueness being removed. Further,
corresponding to this change, step 2 of Algorithm IGWF-WCR will be carried out
in this way: given a feasible point $\left\{ S_i^{(n)} \right\}_{i=1}^{K}$, its image under step 2 of Algorithm
IGWF-WCR is a set of points. The feasible set of problem (4.31) is denoted by V_d.
A point in this set is chosen arbitrarily as the next point $\left\{ S_i^{(n+1)} \right\}_{i=1}^{K}$ generated
by Algorithm IGWF-WCR. Thus, Algorithm IGWF-WCR can generate a point
sequence under this change. In the following, we will still call this algorithm
Algorithm IGWF-WCR despite the changes.

For any convergent subsequence, whose limit is denoted by $(\overline{S}_1, \cdots, \overline{S}_K)$, generated by Algorithm IGWF-WCR, we may use the following lemma to prove that the
limit is a fixed point under Algorithm IGWF-WCR, when Algorithm IGWF-WCR
is regarded as a mapping.

Lemma 4.1. *A point is the limit of a convergent subsequence of the point sequence
generated by Algorithm IGWF-WCR if and only if this point is a fixed point under
Algorithm IGWF-WCR.*

Proof. See [26].

Lemma 4.2. $(\overline{S}_1, \cdots, \overline{S}_K) \in V_d$ *is a fixed point under Algorithm IGWF-WCR if and
only if* $(\overline{S}_1, \cdots, \overline{S}_K) \in V_d$ *is one of the optimal solutions to the problem in* (4.3).

Proof. See [26].

Based on the lemmas above, we obtain the conclusion that Algorithm IGWF-
WCR is convergent. At the same time, step 3 of Algorithm IGWF-WCR is then
regarded as a computation for a point. With these lines of proofs, Algorithm IGWF-
WCR generates a point sequence and every point of the point sequence consists of
the K non-negative numbers, e.g. $\left(S_1^{(n)}, \cdots, S_K^{(n)} \right)$. The details are described below.

Theorem 4.1. *Algorithm IGWF-WCR is convergent. At the same time, the sequence of objective values, obtained by evaluating the objective function at the point sequence, monotonically increases to the optimal objective value.*

Proof. Due to compactness of the set of feasible solutions for the problem in (4.3), the point sequence generated by Algorithm IGWF-WCR already includes a convergent subsequence. For every convergent subsequence, according to Lemma 4.1, the convergent subsequence must converge to a fixed point under Algorithm IGWF-WCR. Then, according to Lemma 4.2, it converges to one of the optimal solutions to the problem in (4.3).

In addition, conversely, as stated by the sufficient and necessary conditions of Lemmas 4.1 and 4.2, for any optimal solution to the problem in (4.3), there is a point sequence generated by Algorithm IGWF-WCR such that the point sequence converges to that optimal solution.

With Algorithm IGWF-WCR generating the point sequence, the definition of Algorithm IGWF-WCR implies that the sequence of the objective values, obtained by evaluating the objective function at the point sequence, monotonically increases to the optimal objective value. This is due to the monotonicity and any convergent subsequence of the point sequence converging to one of the optimal solutions to the optimization problem in (4.3).

Therefore, Algorithm IGWF-WCR is convergent.

To reduce the cost of computation, (4.21) and (4.29) in Sect. 4.3 may utilize the Fibonacci search. To improve the performance of the algorithm and reduce the cost of the computation, the objective function in step 3 of the IGWF-WCR can be evaluated at the four points mentioned in Remark 4.3, by parallel computation to find the estimate of β^* of (4.29).

4.5   Numerical Examples for RRM in Cognitive Radio Network

In this section, numerical examples are provided to illustrate the effectiveness of the proposed algorithm IGWF-WCR. For comparison purpose, a regular feasible direction method utilizing the gradient [27] in the optimization is chosen. It is denoted as algorithm AFD. Note that, as a benchmark and a feasible direction method, the algorithm AFD can also generate a sequence of feasible points (as a feasible point algorithm). It is easy to set up a stop criterion of computation for a feasible point algorithm, especially for a monotonic feasible point algorithm like the proposed one. Due to the feasible set being a convex polygon, the recently developed AFD algorithm is used as a reference. We didn't select [18] for comparison since the primal-dual algorithm used in [18] is not a feasible point method; in addition, the assumption of the constrains is different and system model is different, too.

Figures 4.2 and 4.4 show the evolution of the weighted sum-rate values versus the number of iterations for IGWF-WCR and AFD for some choices of the number of users (K). In the calculation, the number of antennas at the base station (m) is set to be 4. Channel gain vectors are generated randomly using random $m \times 1$ vectors with each entry drawn independently from the standard Gaussian distribution. $\{P_k\}$ is the set of randomly chosen positive numbers. The sum power constraint is $P_t = 10\,\mathrm{dB}$. A group of different weights are also generated randomly. In these figures, the cross markers and the diamond markers represent the results of our proposed algorithm IGWF-WCR and the AFD respectively. These results show that the proposed algorithm IGWF-WCR exhibits much faster convergence rate, especially with an increasing number of users.

Let f^* be the maximum sum-rate, $f^{(n)}$ the sum-rate at the n-th iteration and $|f^{(n)} - f^*|$ the error in the sum-rate. Figures 4.3 and 4.5 show the corresponding error in the sum-rate versus the number of iterations. Note it is easy to see that using the fixed-point theory of the proposed Lemma 4.2 one can determine the maximum sum-rate f^* mentioned. As shown in these figures, the algorithms converge linearly. The proposed algorithm exhibits a much larger slope in the sum-rate error function, which translates to a faster convergence rate.

Note that the content of this chapter comes partially from [26] and references therein.

References

1. H. Jiang, L. Lai, R. Fan and H. V. Poor, "Optimal selection of channel sensing order in cognitive radio," IEEE Transactions on Wireless Communications, vol. 8, pp. 297–307, 2009.
2. J. Mitola and G. Q. Maguire, "Cognitive radios: Making software radios more personal," IEEE Personal Communications, vol. 6, pp. 13–18, 1999.
3. S. Haykin, "Cognitive radio: Brain-empowered wireless communications," IEEE Journal of Selected Areas in Communications, vol. 23, pp. 201–220, 2005.
4. R.V. Prasad, P. Pawelczak, J.A. Hoffmeyer and H.S. Berger, "Cognitive functionality in next generation wireless networks: standardization efforts," IEEE Communications Magazine, vol. 46, pp. 72–78, 2008.
5. N. Devroye, M. Vu and V. Tarokh, "Cognitive radio networks: Information theory limits, models and design," IEEE Signal Processing Magazine, vol. 25, pp. 12–23, 2008.
6. D. Tse and S. Hanly, "Multiaccess fading channels. Part I: Polymatroid structure, optimal resource allocation and throughput capacities," IEEE Transactions on Information Theory, vol. 44, pp. 2796–2815, 1998.
7. S. Vishwanath, N. Jindal and A. Goldsmith, "Duality, achievable rates and sum-rate capacity of Gaussian MIMO broadcast channels," IEEE Transactions on Information Theory, vol. 49, pp. 2658–2668, 2003.
8. E. Biglieri, R. Calderbank, A. Constantinides, A. Goldsmith, A. Paulraj and H. V. Poor, MIMO Wireless Communications, Cambridge University Press, Cambridge, 2007.
9. N. Jindal, W. Rhee, S. Vishwanath, S. A. Jafar and A. Goldsmith, "Sum power iterative water-filling for multi-antenna Gaussian broadcast channels," IEEE Transactions on Information Theory, vol. 51, pp. 1570-=1580, 2005.
10. W. Yu, "Sum-capacity computation for the Gaussian vector broadcast channel via dual decomposition," IEEE Transactions on Information Theory, vol. 52, pp. 754–759, 2006.
11. W. Yu, W. Rhee, S. Boyd and J. M.Cioffi, "Iterative water-filling for Gaussian vector multi-access channels," IEEE Transactions on Information Theory, vol. 50, pp. 145–152, 2004.

12. E. Telatar, "Capacity of multi-antenna Gaussian channels," European Transactions on Telecommunications, vol. 10, pp. 585–596, 1999.

13. N. Jindal, S. Vishwanath and A. Goldsmith, "On the duality of Gaussian multiple-access and broadcast channels," IEEE Transactions on Information Theory, vol. 50, pp. 768–783, 2004.

14. P. Viswanath and D. Tse, "Sum capacity of the multiple antenna Gaussian broadcast channel and uplink-downlink duality," IEEE Transactions on Information Theory, vol. 49, pp. 1912–1921, 2003.

15. H. Weingarten, Y. Steinberg and S. Shamai, "The capacity region of the Gaussian multiple-input multiple-output broadcast channel," IEEE Transactions on Information Theory, vol. 52, pp. 3936–3964, 2006.

16. M. Kobayashi and G. Caire, "An iterative water-filling algorithm for maximum weighted sum-rate of Gaussian MIMO-BC," IEEE Journal of Selected Areas in Communications, vol. 24, pp. 1640–1646, 2006.

17. L. Zhang, Y.-C. Liang and Y. Xin, "Joint beamforming and power allocation for multiple access channels in cognitive radio networks," IEEE Journal of Selected Areas in Communications, vol. 26, pp. 38–51, 2008.

18. L. Zhang, Y. Xin, Y.-C. Liang and H. V. Poor, "Cognitive multiple access channels: Optimal power allocation for weighted sum rate maximization," IEEE Transactions on Communications, vol. 57, pp. 2754–2762, 2009.

19. R. Zhang, S. Cui and Y. C. Liang, "On ergodic sum capacity of fading cognitive multiple-access and broadcast channels," IEEE Transactions on Information Theory, vol. 55, pp. 5161–5178, 2009.

20. D. Palomar, "Practical algorithms for a family of waterfilling solutions," IEEE Transactions on Signal Processing, vol. 53, pp. 686–695, 2005.

21. C. Hs, H. Su and P. Lin, "Joint subcarrier pairing and power allocation for OFDM transmission with decode-and-forward relaying," IEEE Transactions on Information Theory, vol. 59, pp. 399–414, 2011.

22. Q. Qi, A. Minturn and Y. Yang, "An efficient water-filling algorithm for power allocation in OFDM-based cognitive radio systems," Systems and Informatics (ICSAI), 2012 International Conference on, Yantai, pp. 2069–2073, 2012.

23. Y. Rong, X. Tang and Y. Hua, "A unified framework for optimizing linear non-regenerative multicarrier MIMO relay communication systems," IEEE Transactions on Signal Processing, vol. 57, pp. 4837–4852, 2009.

24. A. Quarteroni, R. Sacco and F. Saleri, Numerical Mathematics, 2nd Edition, Springer Berlin Heidelberg, 2010.

25. W. Zangwill, Nonlinear Programming: A Unified Approach, Prentice-Hall, Englewood Cliffs, 1969.

26. P. He, L.Zhao and J. H. Lu, "Weighted sum-rate maximization for multi-user SIMO multiple access channels in cognitive radio networks," EURASIP Journal on Advances in Signal Processing, 15 pages, doi: 10.1186/1687-6180-2013-80, April, 2013.

27. W. Sun and Y. Yuan, Optimization Theory and Methods: Nonlinear Programming, 1st Edition, (Springer Optimization and Its Applications), Springer, 2006.

Chapter 5
RRM in Wireless Communications with Energy Harvest Technology

5.1 Introduction

There has been recent research effort on understanding data transmission with an energy harvesting transmitter that has a rechargeable battery for green communications [1–4]. Recently, based on [3, 5] further investigates the issues of power allocation problems to minimize the grid power consumption with random energy and data arrival. In more detail, for the implied problem that is a convex optimization problem, rather than the original non-convex problem, a solution is computed. For convenience and without loss of generality, the process is considered as a discrete time process. The simplest and useful system model, illustrated in Fig. 5.1, assumes that there are K epochs in the time period $(0, T]$. For each epoch, an event occurs which may be the consequence of channel fading gain variation or new energy arrival, or both. This setting leads to new design insights in a wireless link with a rechargeable transmitter and fading channels.

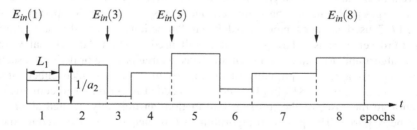

Fig. 5.1 Illustration of simple system instance

Besides the allocated power to be non-negative, sum of successively harvested energy over time determines a more complicated power constraint. Generally, the incoming energy can be stored in the battery of the rechargeable transmitter for future use. However, it cannot be used before its arrival. This point is called

P. He et al., *Radio Resource Management Using Geometric Water-Filling*, SpringerBriefs in Computer Science, DOI 10.1007/978-3-319-04636-5_5, © The Author(s) 2014

causality. Often, this battery owns a great storage capacity and it is hardly filled fully from the incoming harvested energy. This assumption is taken into account in this chapter, i.e., the maximum energy capacity of the battery, $E_{max} \gg 0$. This assumption lays down a solid foundation to solve the cases of finite E_{max}. In this setting, we can compute optimal transmission schemes that adapt the instantaneous transmit power to the variations in the energy and fading levels. Since the proposed optimal dynamic transmission power allocation policy results from the recursive computing, which does not utilize any information in time future, the optimal dynamic power allocation can provide the optimal solution to the maximum throughput for every sub-process or time window from epoch 1 until epoch k, as $k = 1, \ldots, K$. This advantage owns more challenge and could be utilized to efficiently solve other problems.

In recent years, energy harvesting green communication has attracted great research attention. In [1], data transmission with energy harvesting sensors is considered, and the optimal online policy for controlling admissions into the data buffer is derived using a dynamic programming framework. Dynamic programming can offer a real-time feedback policy or control, i.e., considering the time constraint of decisions, but it needs to store a family of offline policies. Further, its object is a dynamic optimization problem model, including the dynamic state transition equation(s). To avoid the curse of dimensionality, a "good" dynamic optimization problem model should be needed or set up. In [2], energy management policies stabilizing the data queue are proposed for single-user communication and under a linear approximation, some delay optimality properties are derived. In [6], the optimality of a variant of the back pressure algorithm using energy queues is shown. In [3] and references therein, optimization approaches are considered to attempt to obtain the maximum throughput over AWGN and fading channels. Successively, in [4], throughput optimal energy allocation is studied for energy harvesting systems in a time constrained slotted setting. Especially in recent work, [3] and [4] investigated the same objective function, in order to attempt to use offline machinery and conventional water-filling approach which come directly from the KKT conditions of the target problem. For the fading channel cases, [3], at the first paragraph on page 1737, used the four sentences to define its "directional water-filling algorithm". The third sentence is a key point but it still used the term "directional water-filling algorithm" to define the algorithm. Thus, a circular logic or definition seems generated. As a result, optimality of the algorithm is not provided either. Since Algorithm 2, on page 4815 in [4], used its embedded Algorithm 1 to compute the solution to the cases of the full side information, an infinitely iterated algorithm may be required. The proposed algorithm in this chapter could overcome these weaknesses.

With water-filling, more power is allocated to the channels with higher gains to maximize the sum of data rates of all the sub-channels [7]. The conventional way to solve the water-filling problem is to solve the KKT conditions, and then find the water-level(s) and the solutions. In this chapter, we exploit our proposed GWF presented in Chap. 2, and construct a recursive algorithm to solve the target problem and then prove its optimality, referred as RGWF. We have shown that GWF owns less computation. This advantage becomes more significant, especially when GWF is utilized multiple times.

Compared with the existing results on energy harvest, the proposed RGWF owns three distinct characteristics: (1) for the fading channel cases, the algorithm is clearly defined; (2) it provides the exact optimal solution via finite computation recursively; (3) its optimality is strictly proven. Therefore, following the proposed algorithm, exact solution can be obtained for any sub-process from time $(0, T]$. Numerical examples provide detailed procedures for determining the optimal solution by the proposed RGWF. Part of the contents in this chapter comes from [8] and the references therein.

In the remaining of the chapter, the proposed GWF is discussed in Sect. 5.2 with sum power constraint. The proposed power allocation problem and RGWF is further investigated in Sect. 5.3. Numerical examples are presented for the proposed RGWF.

5.2 Extended GWF for Transmission with Energy Harvest

Let L and K be two positive integers and $L \leq K$ to denote the index of the starting channel and the ending channel respectively. Often, L is assigned to be 1. The water-filling problem can be abstracted and generalized into the following problem: given $P > 0$, as the total power or volume of the water; the allocated power and the propagation path gain for the ith channel are given as s_i and a_i respectively, $i = L, \ldots, K$; and $K - L + 1$ is the total number of channels. Furthermore, the weighted coefficient $w_i > 0, \forall i$, and $\{a_i w_i\}_{i=L}^{K}$ being monotonically decreasing, find that

$$\begin{aligned} \max_{\{s_i\}_{i=L}^{K}} \quad & \sum_{i=L}^{K} w_i \log(1 + a_i s_i) \\ \text{subject to} : \quad & 0 \leq s_i, \ \forall i; \\ & \sum_{i=L}^{K} s_i = P. \end{aligned} \tag{5.1}$$

Since the constraints are that (i) the allocated power to be nonnegative; (ii) the sum of the power equals P, the problem (5.1) is called the water-filling (problem) with sum power constraint.

In this chapter, we propose a novel approach to solve problem (5.1) based on geometric view. The proposed Geometric Water-Filling (GWF) approach eliminates the procedure to solve the non-linear system for the water level, and provides explicit solutions and helpful insights to the problem and the solution.

Similarly, Fig. 5.2a–d give an illustration of the proposed GWF algorithm. Suppose there are four steps/stairs ($L = 1, K = 4$) inside a water tank. For the conventional approach, the dashed horizontal line, which is the water level μ, needs to be determined first and then the power allocated (water volume) above the step is solved.

Let us use d_i to denote the "step depth" of the ith stair which is the height of the ith step to the bottom of the tank, and is given as

$$d_i = \frac{1}{a_i w_i}, \text{ for } i = L, L+1, \dots, K. \tag{5.2}$$

Since the sequence $a_i w_i$ is sorted as monotonically decreasing, the step depth of the stairs indexed as $\{L, \cdots, K\}$ is monotonically increasing.

Instead of trying to determine the water level μ, which is a real nonnegative number, we aim to determine water level step, which is an integer number from L to K, denoted by k^*, as the highest step under water. Based on the result of k^*, we can write out the solutions for power allocation instantly.

Figure 5.1a illustrates the concept of k^*. Since the third level is the highest level under water, we have $k^* = 3$. The shadowed area denotes the allocated power for the third step by s_3^* (Fig. 5.2).

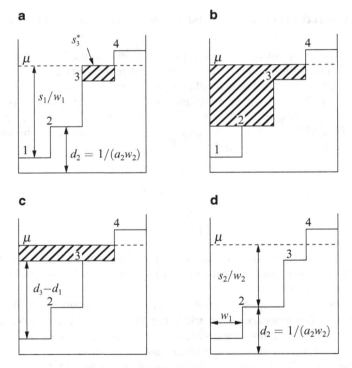

Fig. 5.2 Illustration for the proposed Geometric Water-Filling (GWF) algorithm. (**a**) Illustration of water level step $k^* = 3$, allocated power for the third step s_3^*, and step/stair depth $d_i = \frac{1}{a_i w_i}$. (**b**) Illustration of $P_2(k)$ (*shadowed area*, representing the total water/power above step k) when $k = 2$. (**c**) Illustration of $P_2(k)$ when $k = 3$. (**d**) Illustration of the weights as the widths

In the following, $P_2(k)$, the water volume above step k, can be obtained considering the step depth difference and the width of the stairs as,

$$P_2(k) = \left[P - \sum_{i=L}^{k-1}(d_k - d_i)w_i\right]^+, \text{ for }$$
$$k = L, \ldots, K.$$
(5.3)

where $(x)^+ = \max\{0, x\}$.

As an example in Fig. 5.1c, the water volume above step 1 and below step 3 with the width w_1 can be found as: the step depth difference, $(d_3 - d_1)$ multiplying the width of the step, w_1. Therefore, the corresponding $P_2(k = 3)$ can be expressed as,

$$P_2(k = 3) = [P - (d_3 - d_1)w_1 - (d_3 - d_2)w_2]^+,$$

which is an expansion of (5.3). Then we have the following proposition for integrity.

Proposition 5.1. *The explicit solution to (5.1) is:*

$$\begin{cases} s_i = \left[\dfrac{s_{k^*}}{w_{k^*}} + (d_{k^*} - d_i)\right]w_i, \ L \le i \le k^* \\ s_i = 0, \ k^* < i \le K, \end{cases}$$
(5.4)

where

$$k^* = \max\left\{k \middle| P_2(k) > 0, \ L \le k \le K\right\}$$
(5.5)

and the power level for this step is

$$s_{k^*} = \frac{w_{k^*}}{\sum_{i=L}^{k^*} w_i} P_2(k^*).$$
(5.6)

Proposition 5.1 has been proven above.

GWF can be regarded as a mapping from the point of parameters

$$\{L, K, \{w_i\}_{i=L}^K, \{a_i\}_{i=L}^K, P\}$$

to the solution $\{s_i\}_{i=L}^K$ and the important water level step index: k^*. That is to say, it can be written as a formal expression:

$$\{\{s_i\}_{i=L}^K, k^*\} = GWF(L, K, \{w_i\}_{i=L}^K, \{a_i\}_{i=L}^K, P).$$

Note that, for concision and without confusion from context, we may write the right hand side of the expression, mentioned above, as GWF(L, K) to emphasize time stages from L to K.

5.3 Maximizing Throughput in Fading Channel and Algorithm RGWF

In this section, we firstly introduce the maximizing throughput problem in fading channel and the conventional approach from its KKT conditions. Then, we present the proposed online geometric algorithm.

5.3.1 Problem Statement and Conventional Approach

As shown in Fig. 5.1, we consider the time period from $(0, T]$. The channel state changes or/and energy arrives K times in this time period. Hence, we have K epochs, with L_i being the time length of the ith epoch. At the beginning of the ith epoch, the fading gain is denoted as a_i. The energy arrival is denoted as $E_{in}(i)$, and the event of energy arrival is depicted by $E_{in}(i) \geq 0$. Hence, $E_{in}(i) = E_j$ for some j if event i is an energy arrival and $E_{in}(i) = 0$ if event i is only a fading level change. Our objective is to maximize the number of bits transmitted by the deadline T. The optimal power management strategy is such that the transmit power is constant in each event epoch. Therefore, let us again denote the transmit power in epoch i by s_i, for $i = 1, \ldots, K$.

We have causality constraints due to energy arrivals and an E_{max} constraint due to finite battery size. Hence, the optimization problem in this fading case becomes [3]:

$$\max_{\{s_i\}_{i=1}^K} \quad \sum_{i=1}^K \frac{L_i}{2} \log(1 + a_i s_i)$$
$$\text{subject to: } 0 \leq s_i, \; \forall i;$$
$$\sum_{i=1}^l L_i s_i \leq \sum_{i=1}^l E_{in}(i), \text{ as } l = 1, \ldots, K; \tag{5.7}$$
$$\sum_{i=1}^l E_{in}(i) - \sum_{i=1}^l L_i s_i \leq E_{max}, \forall l,$$

where if we interpret the observed properties of the optimal power allocation scheme as a water-filling scheme mentioned above, $E_{in}(i)$ units of water is filled into a rectangle of bottom width $L_i, \forall i$. With the assumption of $E_{max} \gg 0$, the last constraint of (5.7) disappears. Note that the last power sum constraint in this narrowed problem is of equality. Furthermore, for unifying parameter notation, through a change of variables, we can obtain an equivalent problem to the narrowed one by $E_{max} \gg 0$ as follows:

$$\max_{\{s_i\}_{i=1}^K} \quad \sum_{i=1}^K w_i \log(1 + a_i s_i)$$
$$\text{subject to: } 0 \leq s_i, \; \forall i; \tag{5.8}$$
$$\sum_{i=1}^l s_i \leq \sum_{i=1}^l E_{in}(i), \forall l,$$

where $w_i <= \frac{L_i}{2}, a_i <= \frac{a_i}{L_i}$ and $s_i <= L_i s_i$. Note that the symbol $<=$ has been used at last section as assignment operator.

To find the solution to problem (5.8), the conventional water-filling approach usually starts from the Karush-Kuhn-Tucker (KKT) conditions of the problem as a group of the optimality conditions, then the following system in the variables $\{s_i\}$ and the dual variables can be written as

$$
\begin{cases}
s_i = \left(\frac{w_i}{\sum_{j=i}^{K} \lambda_j} - \frac{1}{a_i} \right)^+ , \text{ for } i = 1, \ldots, K, \\
\sum_{i=1}^{l} w_i \left(\frac{1}{\sum_{j=i}^{K} \lambda_j} - \frac{1}{a_i w_i} \right)^+ \leq \sum_{i=1}^{l} E_{in}(l), \forall l \\
\lambda_j \geq 0, \forall j
\end{cases}
\tag{5.9}
$$

where λ_j is the dual variable corresponding to the jth sum power constraint, for any j. The solution to (5.9) is the solution of the problem (5.8). However, it is not easy to solve (5.9).

5.3.2 Recursive Geometric Water-Filling and Its Optimality

In this section, we propose a novel approach to solve problem (5.8) using our proposed GWF approach. The constraint in (5.8) can be expanded into a matrix form as

$$
\begin{pmatrix}
1 & & & \\
1 & 1 & & \\
 & \cdots & & \\
1 & 1 & \cdots & 1
\end{pmatrix}
\begin{pmatrix}
s_1 \\
s_2 \\
\vdots \\
s_K
\end{pmatrix}
\leq
\begin{pmatrix}
\sum_{i=1}^{1} E_{in}(i) \\
\sum_{i=1}^{2} E_{in}(i) \\
\vdots \\
\sum_{i=1}^{K} E_{in}(i)
\end{pmatrix}.
\tag{5.10}
$$

Since the coefficients matrix forms a triangle matrix, we name the proposed algorithm as Geometric Water-Filling for Triangle Coefficient Matrix (RGWF).

The proposed RGWF(K) is stated as in the following Algorithm description:

Algorithm 1 Pseudocode for RGWF

1: Initialize: $L = 1, K, E_{in}(1), w_1, a_1$;
2: Output the result for epoch 1:
    RGWF(1) $= s_1^* = E_{in}(1)$;
3: **for** $L = 2 : 1 : K$ **do**
4:    Input: $\{E_{in}(L), w_L, a_L\}$;
5:    $\{s_k'\}_{k=1}^{L-1} = $ RGWF$(L-1)$;
6:    **for** $n = L : -1 : 1$ **do**
7:      $W = \{w_j\}_{j=n}^{L}$;
8:      $A = \{a_j\}_{j=n}^{L}$;
9:      $S_T = \sum_{j=n}^{L-1} s_j' + E_{in}(L)$;
10:     $\{\{s_k^*\}_{k=n}^{L}, k^*\} = $ GWF(n, L, W, A, S_T);
11:     **if** $\frac{1}{a_{k^*} w_{k^*}} + \frac{s_{k^*}}{w_k} \geq \frac{1}{a_{n_e} w_{n_e}} + \frac{s_{n_e}'}{w_{n_e}}$ where $n_e = \max\{s_k > 0 | 1 \leq k \leq n-1\}$, or $n == 1$ **then**
12:        Output: RGWF$(L) = \{s_1', \ldots, s_{n-1}', s_n^*, \ldots, s_L^*\}$,
13:        Move to next epoch, i.e., go to Line 16;
14:     **end if**
15:    **end for**
16: **end for**

Fig. 5.3 Illustration for Algorithm RGWF (Line 6–15 for $L = 6$), harvested energy having been allocated up to epoch 5; *horizontal-wave shadowed areas* denote power allocation for the processing window; (**a**) $n = 6, k_e^* = 5$; (**b**) $n = 5, k_e^* = 4$; (**c**) $n = 4, k_e^* = 3$; (**d**) $n = 3, k_e^* = 2$

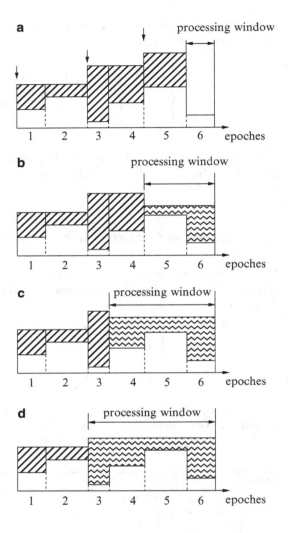

RGWF is illustrated as follows. Based on Lines 1–2 of RGWF, as the base case of the recursive definition, the inner loop (Lines 6–15) can be illustrated in Fig. 5.3 where it is assumed that the current processing epoch $L = 6$. The optimal power allocation for the first five epochs has been completed as shown in the shadowed area in Fig. 5.3a. Epoch 6 is now under processing. Based on Line 9, since there is no harvested energy input in epoch 6, the power level for epoch 6 is zero and the water level is just the fading level. Line 10 calculates that $k_e^* = 5$ and then Line 11 compares the water level of current processing window with that of k_e^*th epoch. Since the comparison in Line 11 does not hold, the algorithm goes back to Line 6 by decreasing n to 5 and then the processing window is extended to include epochs 5 and 6 as shown in Fig. 5.3b. Figure 5.3b also shows the power allocation from

GWF(5,6) in Line 9 as the horizontal-wave shadowed areas. Still, the comparison of the water level non-decreasing in Line 11 does not hold, the algorithm returns to Line 6 again by decreasing $n = 4$. As shown in Fig. 5.3c, the processing window is epochs 4 to 6. The water level non-decreasing condition still is not satisfied. The processing window is extended from epochs 3 to 6 as shown in Fig. 5.3d. With the new water level in the processing window, the water level non-decreasing condition up to epoch 6 is satisfied. As a result, RGWF$(L = 6)$ is solved which is recursively obtained from RGWF$(L - 1 = 5)$ as illustrated in Fig. 5.3.

A summation is used in Line 8. If the lower limit of the summation is greater than the upper limit, the result of this summation is defined as zero, as well known. Through this mechanism, the solution $\{s_i^*\}_{i=1}^K$ is obtained as RGWF(K) within finite loops.

The proposed algorithm eliminates the procedure to solve the non-linear system (5.9) in multiple variables and dual variables, provides online and exact solutions via finite computation steps, and offers helpful insights to the problem and the solutions. To guarantee optimality of RGWF, we have the following proposition:

Proposition 5.2. *RGWF can compute the optimal exact solution to problem (5.8) within finite loops.*

Proof of Proposition 5.2. From the algorithm RGWF(K), applying the KKT conditions can obtain the proof of optimality for RGWF.

Remark 5.1. RGWF is a recursive algorithm with the characteristics of optimal dynamic online power distribution. Dynamics of this algorithm but that for the target problem is shown by the generalized varying structure state equation on dynamics:

$$\text{RGWF}(L+1) = [[\text{RGWF}(L)]|_{\Lambda_1}, [\text{GWF}(n, L+1)]|_{\Lambda_2}], \quad (5.11)$$
$$\text{for } L = 1, \ldots, K - 1,$$

where n is the index of the starting epoch of the currently processing window (i.e., it satisfies Line 11 of RGWF), the set Λ_1 denotes $\{s_k'\}_{k=1}^{n-1}$ and this set is referred to in RGWF mentioned above, and the set Λ_2 denotes $\{s_k\}_{k=n}^{L+1}$ and this set is referred to in Algorithm 1 mentioned before. Thus, we extend the concept of algorithm [9], as a static mapping to a dynamic mapping as a new concept of algorithm to efficiently solve the problems. In this process, RGWF(L) can be regarded as the generalized system state at the time stage (or epoch) L; GWF$(n, L+1)$ can be regarded as the generalized system control at the time stage (or epoch) L; and then RGWF$(L+1)$, as a state at the next time stage, can be derived or determined from the previous state and control. Due to the optimality of RGWF(L) from Proposition 5.2 for any L, the proposed algorithm is indeed an optimal dynamic water-filling algorithm with high efficiency.

5.4   Numerical Examples for RRM with Energy Harvest Transmission

The system model is static. However, our computation does not wait full information input but it can compute the exact optimal solution through finite computation for every sub-process that starts from epoch 1 and ends at epoch k, as $k = 1, \ldots, K$, including the entire process. This point can also lean toward designing other efficient algorithms, such as the algorithm to compute the minimum transmission completion time to avoid a tedious and huge backlog of offline or static computation. The minimum transmission completion time problem can be refereed to in [3] for details. Due to the limit of pages, its discussion is omitted.

For simple illustration, we assume only three epochs, each with unit weight ($w_i = 1, i = 1, 2, 3$). At the beginning of each epoch, unit energy is harvested ($E_{in}(i) = 1, i = 1, 2, 3$).

Example 5.1.   Suppose the fading profile for the three epochs is $a_1 = 1, a_2 = \frac{1}{2}$ and $a_3 = \frac{1}{3}$.

Epoch 1 is first scanned to output RGWF(1)=$s_1 = 1$ as shown in Fig. 5.4a. Now we move to epoch 2 and apply GWF(2,2) and output $s_2 = 1$. Check if the water level of epoch 2 ($2 + 1 = 3$) is greater than the water level of epoch 1 ($1 + 1 = 2$). It is true then output the optimal solution at epoch 2: $s_1 = 1; s_2 = 1$ as shown in Fig. 5.4b. Similarly, for epoch 3, by applying GWF(3,3), we have $s_3 = 1$. Check the water level, it satisfies non-decreasing condition. So the algorithm outputs the completed solution as shown in Fig. 5.4c.

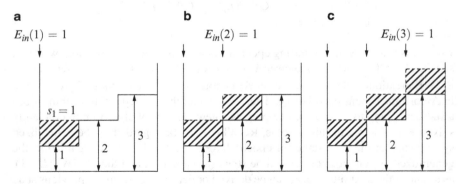

Fig. 5.4   Procedures to solve Example 5.1: (**a**) $s_1 = 1$; (**b**) $s_1 = 1, s_2 = 1$; (**c**) $s_1 = 1, s_2 = 1, s_3 = 1$

Example 5.1 is calculated out without power level adjustment. In the following example, we illustrate the power level adjustment procedure.

Example 5.2.   Suppose the fading profile for the three epochs is $a_1 = 1, a_2 = 2$ and $a_3 = 3$. For this example, the proposed RGWF is illustrated in the following Fig. 5.5.

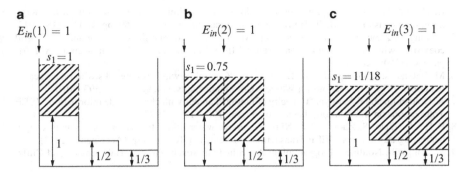

Fig. 5.5 Procedures to solve Example 5.2: (**a**) $s_1 = 1$; (**b**) $s_1 = 0.75, s_2 = 1.25$; (**c**) $s_1 = 11/18, s_2 = 20/18, s_3 = 23/18$

First, we scan the first epoch and RGWF(1) output $s_1 = 1$, as shown in Fig. 5.5a. Then move to the second epoch, by applying GWF(2,2), it gives $s_2 = 1$. Now check the water level of epoch 2 is $1 + 1/2 = 1.5$ and the water level for epoch 1 is $1 + 1 = 2$. Water level non-decreasing condition is violated. Power level adjustment procedure is triggered. By applying GWF to the first two epochs, we have GWF(1,2) $= \{s_1 = 0.75, s_2 = 1.25\}$. With this power adjustment, the new water level for both epochs is 1.75, satisfying non-decreasing condition. The output for RGWF(2) is then $s_1 = 0.75, s_2 = 1.25$ as shown in Fig. 5.5b.

Now we move to epoch 3, the output of GWF(3,3) $= s_3 = 1$. The corresponding water level for epoch 3 is $1 + 1/3$, which is lower than the water level of the previous epoch ($= 1.75$). Then the power adjustment is triggered. The algorithm calculates the power allocation for current epoch (epoch 3) and its previous epoch (epoch 2) to have output GWF(2,3) $= \{s_2 = \frac{25}{24}, s_3 = \frac{29}{24}\}$. We move to water level check step. The new water level of epoch 2 ($\frac{1}{2} + \frac{25}{24} = \frac{37}{24}$) is lower than the water level of epoch 1 (1.75). Therefore, power adjustment needs to include epoch 1 as well. We then compute GWF(1,3), the output is $\{s_1 = \frac{11}{18}, s_2 = \frac{11}{18} + \frac{1}{2}, s_3 = \frac{11}{18} + \frac{2}{3}\}$, which is the completed output for the optimal solution as shown in Fig. 5.5c.

References

1. J. Lei, R. Yates and L. Greenstein, "A generic model for optimizing single-hop transmission policy of replenishable sensors," IEEE Transactions on Wireless Communications, vol. 8, pp. 547–551, 2009.
2. V. Sharma, U. Mukherji, V. Joseph and S. Gupta, "Optimal energy management policies for energy harvesting sensor nodes," IEEE Transactions on Wireless Communications, vol. 9, pp. 1326–1336, 2010.
3. O. Ozel, K. Tutuncuoglu, J. Yang, S. Ulukus and A. Yener, "Transmission with energy harvesting nodes in fading wireless channels: Optimal policies," IEEE Journal of Selected Areas in Communications, vol. 29, pp. 1732–1743, 2011.

4. C. Ho and R. Zhang, "Optimal energy allocation for wireless communications with energy harvesting constraints," IEEE Transactions on Signal Processing, vol. 60, pp. 4808–4818, 2012.
5. J. Gong, S. Zhou and Z. Niu, "Optimal power allocation for energy harvesting and power grid coexisting wireless communication systems," IEEE Transactions on Communications, vol. 61, pp. 3040–3049, 2013.
6. M. Gatzianas, L. Georgiadis and L. Tassiulas, "Control of wireless networks with rechargeable batteries," IEEE Transactions on Wireless Communications, vol. 9, pp. 581–593, 2010.
7. A. Goldsmith and P. Varaiya, "Capacity of fading channels with channel side information," IEEE Transactions on Information Theory, vol. 43, pp. 1986–1992, 1997.
8. P. He, L. Zhao, S. Zhou and Z. Niu, "Recursive water-filling for wireless links with energy harvesting transmitters," IEEE Transactions on Vehicular Technology, (Accepted in 2013).
9. W. Zangwill, Nonlinear Programming: A Unified Approach, Prentice-Hall, Englewood Cliffs, 1969.

Appendix A

A.1 Appendix-I: Complex Gaussian Random Vectors

For any $z \in \mathbb{C}^n$ and $A \in \mathbb{C}^{n \times m}$, let us define

$$\hat{z} = \begin{pmatrix} \mathbb{R}e(z) \\ \mathbb{I}m(z) \end{pmatrix}$$

and

$$\hat{A} = \begin{pmatrix} \mathbb{R}e(A) & -\mathbb{I}m(A) \\ \mathbb{I}m(A) & \mathbb{R}e(A) \end{pmatrix}.$$

A complex random vector $\xi \in \mathbb{C}^n$ is said to be Gaussian if the real random vector $\hat{\xi} \in \mathbb{R}^{2n}$ consisting of the real and imaginary parts of ξ,

$$\hat{\xi} = \begin{pmatrix} \mathbb{R}e(\xi) \\ \mathbb{I}m(\xi) \end{pmatrix},$$

is Gaussian [1]. In fact, any complex random vector, without restricting to a complex Gaussian random vector, has such a real and expanded random vector. The relationship between these two complex and real vectors is a one to one mapping. Let us recall

$$E\left[\hat{\xi}\right] \triangleq \int_{\mathbb{R}^{2n}} x \, dP_{\hat{\xi}}(x) \in \mathbb{R}^{2n}$$

and

$$E\left[\left(\hat{\xi} - E\left[\hat{\xi}\right]\right)\left(\hat{\xi} - E\left[\hat{\xi}\right]\right)^T\right] \triangleq \int_{\mathbb{R}^{2n}} \left(x - E\left[\hat{\xi}\right]\right)\left(x - E\left[\hat{\xi}\right]\right)^T dP_{\hat{\xi}}(x)$$

P. He et al., *Radio Resource Management Using Geometric Water-Filling*, SpringerBriefs in Computer Science, DOI 10.1007/978-3-319-04636-5, © The Author(s) 2014

$\in \mathbb{R}^{2n \times 2n}$, the former is called the mean of $\hat{\xi}$ and the latter is called the covariance of $\hat{\xi}$. Note that $\mathbb{R}^{2n \times 2n}$ stands for the set of real square matrices with the size of $(2n) \times (2n)$, i.e., $2n$ rows and $2n$ columns. According to standard Lebesgue integration [2] on $\mathbb{R}^{2n}$, the mean and covariance of $\hat{\xi}$ can be found respectively. Thus, to specify the distribution of a complex Gaussian random vector ξ, it is necessary to specify the mean and covariance of $\hat{\xi}$, namely,

$$E\left[\hat{\xi}\right] \text{ and } E\left[\left(\hat{\xi} - E\left[\hat{\xi}\right]\right)\left(\hat{\xi} - E\left[\hat{\xi}\right]\right)^T\right].$$

The definitions of the mean and covariance are also suitable for the case of any complex random vector.

According to standard Lebesgue integration on $\mathbb{C}^n$, mean μ and covariance Q of ξ can be defined as follows.

$$\mu = E[\xi], \text{ where } E[\xi] \triangleq \int_{\mathbb{C}^n} x \, dP_\xi(x) \in \mathbb{C}^n.$$

$$Q = E\left[(\xi - \mu)(\xi - \mu)^\dagger\right],$$

where

$$E\left[(\xi - \mu)(\xi - \mu)^\dagger\right] \triangleq \int_{\mathbb{C}^n} (x - \mu)(x - \mu)^\dagger \, dP_\xi(x) \in \mathbb{C}^{n \times n}.$$

A complex Gaussian random vector ξ is said to be circularly symmetric if the covariance of the corresponding vector $\hat{\xi}$ has the structure

$$E\left[\left(\hat{\xi} - E\left[\hat{\xi}\right]\right)\left(\hat{\xi} - E\left[\hat{\xi}\right]\right)^T\right] = \frac{1}{2}\begin{pmatrix} \mathbb{R}e(Q) & -\mathbb{I}m(Q) \\ \mathbb{I}m(Q) & \mathbb{R}e(Q) \end{pmatrix} \tag{A.1}$$

for some Hermitian positive semidefinite matrix $Q \in \mathbb{C}^{n \times n}$. Note that the real part of a Hermitian matrix is symmetric, and the imaginary part of a Hermitian matrix is skew-symmetric. Thus the matrix appearing in (A.1) is real and symmetric. In this case $E\left[(\xi - E[\xi])(\xi - E[\xi])^\dagger\right] = Q$, and thus, a circularly symmetric complex Gaussian random vector ξ is specified by its mean and variance.

Let ξ be a circularly symmetric complex Gaussian random vector. Then the probability density function (with respect to the Radon-Nikodym derivative of the standard Lebesgue measure on $\mathbb{C}^n$) of a circularly symmetric complex Gaussian random vector with mean μ and covariance Q is derived by the following:

Due to the definition of ξ and the relationship between $\hat{\xi}$ and ξ,

$$f_{\hat{\xi}}(\hat{x};\hat{\mu},\hat{Q}) = \frac{1}{(2\pi)^{\frac{2n}{2}}\left[\det\left(\frac{1}{2}\hat{Q}\right)\right]^{\frac{1}{2}}}\exp\left\{-\frac{1}{2}(\hat{x}-\hat{\mu})^{\dagger}\left(\frac{1}{2}\hat{Q}\right)^{-1}(\hat{x}-\hat{\mu})\right\}$$

$$= \frac{1}{\pi^{n}\left[\det\left(\hat{Q}\right)\right]^{\frac{1}{2}}}\exp\left\{-(\hat{x}-\hat{\mu})^{\dagger}\left(\hat{Q}\right)^{-1}(\hat{x}-\hat{\mu})\right\}.$$

The next step is to simplify the exponent part from the previous exponent function in order to finally remove the symbol "^". This can be done by utilizing the following algebraic property.

Due to $C = A^{-1}$ being equivalent to $\hat{C} = \left(\hat{A}\right)^{-1}$, we have

$$\frac{1}{\pi^{n}\left[\det\left(\hat{Q}\right)\right]^{\frac{1}{2}}}\exp\left\{-(\hat{x}-\hat{\mu})^{\dagger}\left(\hat{Q}\right)^{-1}(\hat{x}-\hat{\mu})\right\}$$

$$= \frac{1}{\pi^{n}\left[\det\left(\hat{Q}\right)\right]^{\frac{1}{2}}}\exp\left\{-(\hat{x}-\hat{\mu})^{\dagger}\widehat{\left(Q^{-1}\right)}(\hat{x}-\hat{\mu})\right\}.$$

The next step is to remove the symbol "^" in the preceding exponent function. Due to $z = x + y$ being equivalent to $\hat{z} = \hat{x} + \hat{y}$, $\mathbb{R}e\left(x^{\dagger}y\right) = \hat{x}^{\dagger}\hat{y}$ and $y = Ax$ being equivalent to $\hat{y} = \hat{A}\hat{x}$, we have

$$\frac{1}{\pi^{n}\left[\det\left(\hat{Q}\right)\right]^{\frac{1}{2}}}\exp\left\{-(\hat{x}-\hat{\mu})^{\dagger}\widehat{\left(Q^{-1}\right)}(\hat{x}-\hat{\mu})\right\}$$

$$= \frac{1}{\pi^{n}\left[\det\left(\hat{Q}\right)\right]^{\frac{1}{2}}}\exp\left\{-(x-\mu)^{\dagger}Q^{-1}(x-\mu)\right\}.$$

We may remove the symbol "^" in the determinant: Due to $\det\left(\hat{A}\right) = |\det(A)|^{2}$, we have

$$\frac{1}{\pi^{n}\left[\det\left(\hat{Q}\right)\right]^{\frac{1}{2}}}\exp\left\{-(x-\mu)^{\dagger}Q^{-1}(x-\mu)\right\}$$

$$= \frac{1}{\pi^{n}\det\left(Q\right)}\exp\left\{-(x-\mu)^{\dagger}Q^{-1}(x-\mu)\right\}.$$

Therefore,

$$f_{\hat{\xi}}(\hat{x};\hat{\mu},\hat{Q}) = \frac{1}{\pi^{n}\det\left(Q\right)}\exp\left\{-(x-\mu)^{\dagger}Q^{-1}(x-\mu)\right\}$$

$$= \det\left(\pi Q\right)^{-1}\exp\left\{-(x-\mu)^{\dagger}Q^{-1}(x-\mu)\right\}.$$

According to the uniqueness of the Radon-Nikodym derivative and the correspondence between $\mathbb{R}^{2n}$ and $\mathbb{C}^n$, we have

$$f_{\xi}(x;\mu,Q) = \det(\pi Q)^{-1} \exp\left\{-(x-\mu)^{\dagger} Q^{-1} (x-\mu)\right\},$$

as the probability density function of ξ.

Remark A.1. Only due to the uniqueness of the Radon-Nikodym derivative, mentioned above, in measure theory, may we acquire the probability density distribution of the complex random vector.

A.2   Appendix-II: Maximum of Entropy

The channel capacity is dependent on the definition of the mutual information. At the same time the mutual information can be computed also by introducing the differential entropy and the conditional entropy. Therefore, the mutual information, the differential entropy and the conditional entropy are revisited, referring to [3] and reference therein. If familiarity of these concepts are assumed, we may skip them over to (A.2).

Definition A.1 (Mutual Information). Assume that $\xi \in \mathbb{C}^n$ and $\eta \in \mathbb{C}^n$ are two complex continuous random vectors, and $p(x)$ and $p(y)$ are the corresponding probability density functions. The mutual information $\mathbb{I}(\xi;\eta)$ between the two random vectors is defined as:

$$\mathbb{I}(\xi;\eta) \triangleq \int_{\mathbb{C}^n} \int_{\mathbb{C}^n} p(x) p(y|x) \log \frac{p(x) p(y|x)}{p(x) p(y)} dx dy,$$

where $p(y|x)$ denotes the conditional probability density function.

In information theory, the mutual information of two random variables (or vectors) is a quantity that measures the mutual dependence of the two variables (or vectors).

In information theory, the following concept of the differential entropy is measurement for the entropy of a random variable (or vector).

Definition A.2 (Differential Entropy). Assume that ξ is a complex continuous random vector, and $p(x)$ is the corresponding probability density function. Then

$$\mathbb{H}(\xi) \triangleq -\int_{\mathbb{C}^n} p(x) \log p(x) dx$$

is called the differential entropy of ξ.

In information theory, the conditional entropy quantifies the remaining entropy of a random variable (or vector) ξ given that the value of a second random variable (or vector) η is known.

Definition A.3 (Conditional Entropy). Assume that ξ and η are two complex continuous random vectors, $p(x)$ and $p(y)$ are the corresponding probability density functions and $p(x,y)$ is the corresponding joint probability density function. Then the conditional entropy of ξ for given η is defined as:

$$\mathbb{H}(\xi|\eta) \triangleq \int_{\mathbb{C}^n} \int_{\mathbb{C}^n} p(x,y) \log p(x|y) \, dx dy.$$

The following proposition offers the mathematical relationship among the mutual information, the differential entropy and the conditional entropy.

Proposition A.1. *Assume that ξ and η are two continuous random vectors. Then*

$$\mathbb{I}(\xi;\eta) = \mathbb{H}(\xi) - \mathbb{H}(\xi|\eta)$$

and

$$\mathbb{I}(\xi;\eta) = \mathbb{H}(\eta) - \mathbb{H}(\eta|\xi).$$

The differential entropy of a complex Gaussian variable (or vector) ξ with mean μ and covariance Q is derived as follows. Due to the definition of $\mathbb{H}(\xi;\mu,Q)$, we have

$$\mathbb{H}(\xi;\mu,Q) = E_\xi \left[-\log f_\xi(\xi;\mu,Q) \right], \tag{A.2}$$

where E_ξ is the expectation operator of ξ, i.e., $E_\xi[\xi] \triangleq \int_{\mathbb{C}^n} x p_\xi(x) \, dx$.

Due to the form of the probability density function of the circularly symmetric complex Gaussian random vector ξ, we have

$$E_\xi \left[-\log f_\xi(\xi;\mu,Q) \right] = \log \det(\pi Q) + E\left[(\xi - \mu)^\dagger Q^{-1} (\xi - \mu) \right].$$

Due to the definition and basic properties of the trace operator, we may write

$$\log \det(\pi Q) + E\left[(\xi - \mu)^\dagger Q^{-1} (\xi - \mu) \right]$$
$$= \log \det(\pi Q) + E\left[\mathrm{Tr} \left((\xi - \mu)(\xi - \mu)^\dagger Q^{-1} \right) \right].$$

Due to the commutative property for the product of the trace and expectation operators, we have

$$\log\det(\pi Q) + E\left[\operatorname{Tr}\left((\xi-\mu)(\xi-\mu)^{\dagger}Q^{-1}\right)\right]$$
$$= \log\det(\pi Q) + \operatorname{Tr}\left(E\left[(\xi-\mu)(\xi-\mu)^{\dagger}\right]Q^{-1}\right).$$

Then the definition of the covariance of ξ implies that

$$\log\det(\pi Q) + \operatorname{Tr}\left(E\left[(\xi-\mu)(\xi-\mu)^{\dagger}\right]Q^{-1}\right) = \log\det(\pi Q) + \operatorname{Tr}\left(QQ^{-1}\right),$$

and using the definition of the logarithm function and the fact, $e \triangleq \lim_{m\to\infty}\left(1+\frac{1}{m}\right)^{m}$, we have

$$\log\det(\pi Q) + \operatorname{Tr}\left(QQ^{-1}\right) = \log\det(\pi Q) + \log e^{n}.$$

This can be simplified to

$$\log\det(\pi Q) + \log e^{n} = \log\det(\pi e Q).$$

The following proposition, which states that a circularly symmetric complex Gaussian variable (or vector) is the entropy maximizer, highlights the importance of circularly symmetric complex Gaussian vectors. Telatar [1] also claims this proposition. For proving that a circularly symmetric complex Gaussian variable (or vector) is the entropy maximizer, we can use the argument, i.e., $\log\gamma_{Q}(x)$ is a linear combination of the terms $x_{i}x_{j}^{*}$. But it is incorrect and unnecessary. In addition, the last step of deriving $\mathbb{H}(p) - \mathbb{H}(\gamma_{Q}) \leq 0$, and $\mathbb{H}(p) - \mathbb{H}(\gamma_{Q}) = 0$ implying $p = \gamma_{Q}$ are not proved. Thus, we offer a formal and alternative proof.

Proposition A.2. *Suppose that the complex random vector $\xi \in \mathbb{C}^{n}$ has zero mean and ξ satisfies $E\left[\xi\xi^{\dagger}\right] = Q$, i.e., $E\left[\xi_{i}\xi_{j}^{\dagger}\right] = Q_{i,j}, 1 \leq i, j \leq n$. Then the entropy of ξ satisfies $\mathbb{H}(f(\xi;\mu,Q)) \leq \log\det(\pi e Q)$, with equality if and only if ξ is a circularly symmetric complex Gaussian random variable (or vector).*

The following two important facts are needed to complete our proof. *The first important fact is:*

$$E_{\eta}\left[\log p_{\eta}(\eta)\right] = E_{\xi}\left[\log p_{\eta}(\xi)\right].$$

The second one is a simple but crucial inequality in our proof. *The second important fact is:*

$$\log x \leq x - 1, \forall x > 0; \quad \log x = x - 1, \text{ iff } x = 1.$$

The first one needs a proof that is given as follows.

Lemma A.1. *If assumptions are the same as those of Proposition A.2, the random vector ξ and η satisfy the assumptions and η is a circularly symmetric complex Gaussian random vector, then*

$$E_\eta \left[\log p_\eta (\eta)\right] = E_\xi \left[\log p_\eta (\xi)\right].$$

Proof (The Proof of Lemma A.1). Due to the definitions of η and (A.2), which qualify the relationship expression between the differential entropy and the mean, we have

$$E_\eta \left[\log p_\eta (\eta)\right] = \int_{\mathbb{C}^n} \left[-\log \det (\pi Q) - (x - \mu)^\dagger Q^{-1} (x - \mu)\right] p_\eta (x)\, dx.$$

Due to the linearity property of integration, we have

$$\int_{\mathbb{C}^n} \left[-\log \det (\pi Q) - (x - \mu)^\dagger Q^{-1} (x - \mu)\right] p_\eta (x)\, dx$$
$$= -\log \det (\pi Q) - \int_{\mathbb{C}^n} (x - \mu)^\dagger Q^{-1} (x - \mu)\, p_\eta (x)\, dx.$$

Due to the circular invariance property of the trace, $\mathrm{Tr}\,(ABC) = \mathrm{Tr}\,(BCA)$, we have

$$-\log \det (\pi Q) - \int_{\mathbb{C}^n} (x - \mu)^\dagger Q^{-1} (x - \mu)\, p_\eta (x)\, dx$$
$$= -\log \det (\pi Q) - \int_{\mathbb{C}^n} \mathrm{Tr}\left(Q^{-1} (x - \mu)(x - \mu)^\dagger\right) p_\eta (x)\, dx.$$

Because the order of the trace and integration can be interchanged, one has

$$-\log \det (\pi Q) - \int_{\mathbb{C}^n} \mathrm{Tr}\left(Q^{-1} (x - \mu)(x - \mu)^\dagger\right) p_\eta (x)\, dx$$
$$= -\log \det (\pi Q) - \mathrm{Tr}\left(\int_{\mathbb{C}^n} Q^{-1} (x - \mu)(x - \mu)^\dagger\, p_\eta (x)\, dx\right).$$

Due to the linearity property of the integration, we have

$$-\log \det (\pi Q) - \mathrm{Tr}\left(\int_{\mathbb{C}^n} Q^{-1} (x - \mu)(x - \mu)^\dagger\, p_\eta (x)\, dx\right)$$
$$= -\log \det (\pi Q) - \mathrm{Tr}\left(Q^{-1} \int_{\mathbb{C}^n} (x - \mu)(x - \mu)^\dagger\, p_\eta (x)\, dx\right).$$

The assumption that the variances of ξ and η are the same implies

$$Q = \int_{\mathbb{C}^n} (x - \mu)(x - \mu)^\dagger p_\eta(x)\,dx = \int_{\mathbb{C}^n} (x - \mu)(x - \mu)^\dagger p_\xi(x)\,dx,$$

$$-\log\det(\pi Q) - \mathrm{Tr}\left(Q^{-1}\int_{\mathbb{C}^n}(x-\mu)(x-\mu)^\dagger p_\eta(x)\,dx\right)$$

$$= \int_{\mathbb{C}^n} -\log\det(\pi Q)\, p_\xi(x)\,dx + \mathrm{Tr}\left(Q^{-1}\int_{\mathbb{C}^n} -(x-\mu)(x-\mu)^\dagger p_\xi(x)\,dx\right).$$

Because of the basic property, $\log(ab) = \log(a) + \log(b)$, of the logarithm, it is obtained that

$$\int_{\mathbb{C}^n} -\log\det(\pi Q)\, p_\xi(x)\,dx + \mathrm{Tr}\left(Q^{-1}\int_{\mathbb{C}^n} -(x-\mu)(x-\mu)^\dagger p_\xi(x)\,dx\right)$$

$$= \int_{\mathbb{C}^n} \log\left(\det(\pi Q)^{-1}\exp\left\{\mathrm{Tr}\left(-Q^{-1}(x-\mu)(x-\mu)^\dagger\right)\right\}\right) p_\xi(x)\,dx.$$

The circular invariance property, $\mathrm{Tr}(ABC) = \mathrm{Tr}(CAB)$, of the trace implies

$$\int_{\mathbb{C}^n} \log\left(\det(\pi Q)^{-1}\exp\left\{\mathrm{Tr}\left(-Q^{-1}(x-\mu)(x-\mu)^\dagger\right)\right\}\right) p_\xi(x)\,dx$$

$$= \int_{\mathbb{C}^n} \log\left(\det(\pi Q)^{-1}\exp\left\{-(x-\mu)^\dagger Q^{-1}(x-\mu)\right\}\right) p_\xi(x)\,dx.$$

Finally, due to the definition of $E_\xi[\log p_\eta(\xi)]$, i.e.,

$$\int_{\mathbb{C}^n} \log\left(\det(\pi Q)^{-1}\exp\left\{-(x-\mu)^\dagger Q^{-1}(x-\mu)\right\}\right) p_\xi(x)\,dx = E_\xi[\log p_\eta(\xi)],$$

we have $E_\eta[\log p_\eta(\eta)] = E_\xi[\log p_\eta(\xi)]$.

The proof of Proposition A.2 is offered as follows.

Proof. Let $p_\xi : \mathbb{C}^n \longrightarrow \mathbb{R}$ be the probability density function of ξ. According to the assumption of the random vector ξ,

$$E(\xi\xi^\dagger) = \int_{\mathbb{C}^n} xx^\dagger p_\xi(x)\,dx = Q.$$

Let η be a circularly symmetric complex Gaussian variable (or vector) with zero mean and variance $E(\eta\eta^\dagger) = Q$. Let the probability density function of η be $p_\eta(x)$.

The definitions of $\mathbb{H}(\xi)$ and $\mathbb{H}(\eta)$ imply

$$\mathbb{H}(\xi) - \mathbb{H}(\eta) = -E_\xi[\log p_\xi(\xi)] + E_\eta[\log p_\eta(\eta)]. \tag{A.3}$$

The first important fact holding is followed by

$$-E_\xi \left[\log p_\xi (\xi)\right] + E_\eta \left[\log p_\eta (\eta)\right] = -E_\xi \left[\log p_\xi (\xi)\right] + E_\xi \left[\log p_\eta (\xi)\right].$$

Because of the linearity property of the expectation,

$$-E_\xi \left[\log p_\xi (\xi)\right] + E_\xi \left[\log p_\eta (\xi)\right] = E_\xi \left[\log \frac{p_\eta (\xi)}{p_\xi (\xi)}\right].$$

Due to the second important fact holding and basic properties of Lebesgue integration, it is to see

$$E_\xi \left[\log \frac{p_\eta (\xi)}{p_\xi (\xi)}\right] \leq E_\xi \left[\frac{p_\eta (\xi)}{p_\xi (\xi)} - 1\right]. \tag{A.4}$$

The definition of the expectation implies

$$E_\xi \left[\frac{p_\eta (\xi)}{p_\xi (\xi)} - 1\right] = E_\xi \left[\frac{p_\eta (\xi)}{p_\xi (\xi)}\right] - E_\xi [1] = 1 - 1 = 0.$$

Hence, according to (A.3), $\mathbb{H}(\xi) - \mathbb{H}(\eta) \leq 0$.

Therefore, taking note of the second important fact, the entropy of ξ satisfies

$$\mathbb{H}(\xi) = -E_\xi \left[\log p_\xi (\xi)\right] \leq \log \det (\pi e Q),$$

with equality if and only if ξ is a circularly symmetric complex Gaussian random variable (or vector) under the given mean and variance.

Remark A.2. Equation (A.4) is explained as follows. First, define $u \log u|_{u=0} \triangleq \lim_{u \downarrow 0} u \log u = 0$ and $\frac{u}{u}|_{u=0} \triangleq \lim_{u \downarrow 0} \frac{u}{u} = 1$. Second, (A.4) holds because

$$E_\xi \left[\log \frac{p_\eta (\xi)}{p_\xi (\xi)}\right] = \int_{\mathbb{C}^n} \log \left(\frac{p_\eta (x)}{p_\xi (x)}\right) p_\xi (x) \, dx$$

$$= \left(\int_{\{x | p_\xi (x) = 0\}} + \int_{\{x | p_\xi (x) > 0\}}\right) \log \left(\frac{p_\eta (x)}{p_\xi (x)}\right) p_\xi (x) \, dx$$

$$= \int_{\{x | p_\xi (x) = 0\}} 0 \, dx + \int_{\{x | p_\xi (x) > 0\}} \log \left(\frac{p_\eta (x)}{p_\xi (x)}\right) p_\xi (x) \, dx$$

$$\leq \int_{\{x | p_\xi (x) > 0\}} \left(\frac{p_\eta (x)}{p_\xi (x)} - 1\right) p_\xi (x) \, dx$$

$$\leq \int_{\mathbb{C}^n} \frac{p_\eta(x)}{p_\xi(x)} p_\xi(x)\,dx - \left(\int_{\{x|p_\xi(x)=0\}} + \int_{\{x|p_\xi(x)>0\}} \right) p_\xi(x)\,dx$$

$$= \int_{\mathbb{C}^n} \frac{p_\eta(x)}{p_\xi(x)} p_\xi(x)\,dx - \int_{\mathbb{C}^n} p_\xi(x)\,dx$$

$$= E_\xi \left[\frac{p_\eta(\xi)}{p_\xi(\xi)} \right] - E_\xi[1]$$

$$= E_\xi \left[\frac{p_\eta(\xi)}{p_\xi(\xi)} - 1 \right].$$

Using the definition of the mutual information, we have that

$$C(H,P) = \max_{p_x} \{ \mathbb{H}(y) - \mathbb{H}(y|x) \,|\, S \succeq 0, \mathrm{Tr}(S) \leq P \},$$

where the model (3.1) implies

$$\max_{p_x} \{ \mathbb{H}(y) - \mathbb{H}(y|x) \,|\, S \succeq 0, \mathrm{Tr}(S) \leq P \}$$

$$= \max_{p_x} \{ \mathbb{H}(y) - \mathbb{H}(z) \,|\, S \succeq 0, \mathrm{Tr}(S) \leq P \}.$$

The assumptions of z in model (3.1) also imply

$$\max_{p_x} \{ \mathbb{H}(y) - \mathbb{H}(z) \,|\, S \succeq 0, \mathrm{Tr}(S) \leq P \}$$

$$= \max_{p_x} \{ \mathbb{H}(y) \,|\, S \succeq 0, \mathrm{Tr}(S) \leq P \} - \mathbb{H}(z).$$

Note that we may assume that x satisfies $E(x^\dagger x) \leq P$ and is a zero mean random vector. Furthermore for such an x, if x is a zero mean random vector with covariance $E(xx^\dagger) = S$, then y is a zero mean random vector with covariance $E(yy^\dagger) = HSH^\dagger + I_r$, which results from the form of model (3.1) and the linearity of the expectation operation, and by Proposition A.2 among such y vectors the entropy is the largest when y is a circularly symmetric complex Gaussian random vector, which is the case when x is a circularly symmetric complex Gaussian random vector by the two facts at the end part of last section. Thus, we can further restrict our attention to the circularly symmetric complex Gaussian random vector x. In this case the mutual information is given by $\log \left(\det \left(I_r + HSH^\dagger \right) \right)$.

The two facts are claimed as follows. A linear transformation of a circularly symmetric complex Gaussian random vector is a circularly symmetric complex Gaussian random vector. The set of circularly symmetric complex Gaussian random vectors is closed for addition. They are used for calculating the channel capacity in the following section.

A.3 Appendix-III: Proofs of the Lemmas in Sect. 3.1.3

Lemma A.2. *For the channel H, there is a unitary matrix U such that $U^\dagger H^\dagger H U = diag(\lambda_1, \cdots, \lambda_t)$ (the diagonal matrix) and*

$$\max\left\{\log\left(\det\left(I_r + HSH^\dagger\right)\right) \mid S \succeq 0, Tr(S) \le P\right\} =$$
$$\max\left\{\log\left(\det\left(I_t + diag(\lambda_1, \cdots, \lambda_t)S\right)\right) \mid S \succeq 0, Tr(S) \le P\right\}$$

and $U^\dagger S_l U = S_r$, where S_l and S_r are two optimal solutions of the two optimization problems mentioned above, respectively.

Proof. According to the matrix theory, it is easily known that there is a unitary matrix U such that

$$U^\dagger H^\dagger H U = diag(\lambda_1, \cdots, \lambda_t) \tag{A.5}$$

(the diagonal matrix) and the two maximum points exist due to the compactness of the two constraints. Let

$$S_l \in \arg\max\left\{\log\left(\det\left(I_r + HSH^\dagger\right)\right) \mid S \succeq 0, \mathrm{Tr}(S) \le P\right\}.$$

Because U is a unitary matrix and $\det(I + AB) = \det(I + BA)$ with appropriate dimensions of the matrices, we have

$$\log\left(\det\left(I_r + HS_lH^\dagger\right)\right) = \log\left(\det\left(I_t + U^\dagger H^\dagger H U U^\dagger S_l U\right)\right).$$

Due to (A.5),

$$\log\left(\det\left(I_t + U^\dagger H^\dagger H U U^\dagger S_l U\right)\right) = \log\left(\det\left(I_t + diag(\lambda_1, \cdots, \lambda_t) U^\dagger S_l U\right)\right).$$

Since $S_l \succeq 0$ and $\mathrm{Tr}(S_l) \le P$,

$$\log\left(\det\left(I_t + diag(\lambda_1, \cdots, \lambda_t) U^\dagger S_l U\right)\right)$$
$$\le \max\left\{\log\left(\det\left(I_t + diag(\lambda_1, \cdots, \lambda_t) U^\dagger S U\right)\right) \mid U^\dagger S U \succeq 0, \mathrm{Tr}(U^\dagger S U) \le P\right\}.$$

As the unitary similarity transformation keeps the semidefinite positiveness and trace,

$$\max\left\{\log\left(\det\left(I_t + diag(\lambda_1, \cdots, \lambda_t) U^\dagger S U\right)\right) \mid U^\dagger S U \succeq 0, \mathrm{Tr}(U^\dagger S U) \le P\right\}$$
$$\le \max\left\{\log\left(\det\left(I_t + diag(\lambda_1, \cdots, \lambda_t) S\right)\right) \mid S \succeq 0, \mathrm{Tr}(S) \le P\right\}.$$

Hence,

$$\max \left\{ \log \left(\det \left(I_r + HSH^\dagger \right) \right) | S \succeq 0, \text{Tr}\,(S) \le P \right\} \le$$
$$\max \left\{ \log \left(\det \left(I_t + \text{diag}\,(\lambda_1, \cdots, \lambda_t)\,S \right) \right) | S \succeq 0, \text{Tr}\,(S) \le P \right\}.$$

On the other hand,

$$\forall S_r, S_r \in \arg\max \left\{ \log \left(\det \left(I_t + \text{diag}\,(\lambda_1, \cdots, \lambda_t)\,S \right) \right) | S \succeq 0, \text{Tr}\,(S) \le P \right\}.$$

Because the definition of matrix Λ,

$$\log \left(\det \left(I_t + \text{diag}\,(\lambda_1, \cdots, \lambda_t)\,S_r \right) \right) \le \log \left(\det \left(I_t + U^\dagger H^\dagger H U S_r \right) \right).$$

Due to $\det\,(I + AB) = \det\,(I + BA)$,

$$\log \left(\det \left(I_t + U^\dagger H^\dagger H U S_r \right) \right) \le \log \left(\det \left(I_r + H U S_r U^\dagger H^\dagger \right) \right).$$

As the unitary similarity transformation keeps the semidefinite positiveness and trace, we get

$$\log \left(\det \left(I_r + H U S_r U^\dagger H^\dagger \right) \right)$$
$$\le \max \left\{ \log \left(\det \left(I_r + H U S_r U^\dagger H^\dagger \right) \right) | U S_r U^\dagger \succeq 0, \text{Tr}\,\left(U S_r U^\dagger \right) \le P \right\}.$$

The same reason implies

$$\max \left\{ \log \left(\det \left(I_r + H U S_r U^\dagger H^\dagger \right) \right) | U S_r U^\dagger \succeq 0, \text{Tr}\,\left(U S_r U^\dagger \right) \le P \right\}$$
$$\le \max \left\{ \log \left(\det \left(I_r + HSH^\dagger \right) \right) | S \succeq 0, \text{Tr}\,(S) \le P \right\}.$$

Thus,

$$\max \left\{ \log \left(\det \left(I_t + \text{diag}\,(\lambda_1, \cdots, \lambda_t)\,S \right) \right) | S \succeq 0, \text{Tr}\,(S) \le P \right\}$$
$$\le \max \left\{ \log \left(\det \left(I_r + HSH^\dagger \right) \right) | S \succeq 0, \text{Tr}\,(S) \le P \right\}.$$

Therefore,

$$\max \left\{ \log \left(\det \left(I_r + HSH^\dagger \right) \right) | S \succeq 0, \text{Tr}\,(S) \le P \right\}$$
$$= \max \left\{ \log \left(\det \left(I_t + \text{diag}\,(\lambda_1, \cdots, \lambda_t)\,S \right) \right) | S \succeq 0, \text{Tr}\,(S) \le P \right\}$$

and further $U^\dagger S_l U = S_r$, where S_l and S_r are two optimal solutions of the two optimization problems respectively.

Lemma A.3. *For the channel H, there is a unitary matrix U such that $U^\dagger H^\dagger H U = \Lambda$ (the diagonal matrix) and*

$$\max\left\{\log\left(\det\left(I_r + HSH^\dagger\right)\right)|S \succeq 0, Tr(S) \leq P\right\}$$
$$= \max\left\{\log\left(\det\left(I_t + \Lambda^{\frac{1}{2}}S\Lambda^{\frac{1}{2}}\right)\right)|S \succeq 0, Tr(S) \leq P\right\},$$

and $U^\dagger S_l U = S_r$, where S_l and S_r are two optimal solutions of the two optimization problems mentioned above, respectively.

Proof. According to **Lemma A.2**, we have

$$\max\left\{\log\left(\det\left(I_r + HSH^\dagger\right)\right)|S \succeq 0, Tr(S) \leq P\right\}$$
$$= \max\left\{\log\left(\det\left(I_t + \text{diag}\left(\lambda_1, \cdots, \lambda_t\right)S\right)\right)|S \succeq 0, Tr(S) \leq P\right\}.$$

According to the definition of the matrix Λ and (A.5), we have

$$\max\left\{\log\left(\det\left(I_t + \text{diag}\left(\lambda_1, \cdots, \lambda_t\right)S\right)\right)|S \succeq 0, Tr(S) \leq P\right\}$$
$$= \max\left\{\log\left(\det\left(I_t + \Lambda S\right)\right)|S \succeq 0, Tr(S) \leq P\right\}.$$

Due to the definition of the square root for the matrix Λ, we get

$$\max\left\{\log\left(\det\left(I_t + \Lambda S\right)\right)|S \succeq 0, Tr(S) \leq P\right\}$$
$$= \max\left\{\log\left(\det\left(I_t + \Lambda^{\frac{1}{2}}\Lambda^{\frac{1}{2}}S\right)\right)|S \succeq 0, Tr(S) \leq P\right\}.$$

For the reason that $\det(I + AB) = \det(I + BA)$, we have

$$\max\left\{\log\left(\det\left(I_t + \Lambda^{\frac{1}{2}}\Lambda^{\frac{1}{2}}S\right)\right)|S \succeq 0, Tr(S) \leq P\right\}$$
$$= \max\left\{\log\left(\det\left(I_t + \Lambda^{\frac{1}{2}}S\Lambda^{\frac{1}{2}}\right)\right)|S \succeq 0, Tr(S) \leq P\right\}.$$

References

1. E. Telatar, "Capacity of multi-antenna Gaussian channels," European Transactions on Telecommunications, vol. 10, pp. 585–596, 1999.
2. J. L. Doob, Measure Theory (Graduate Texts in Mathematics), Springer-Verlag, New York, 1994.
3. E. Biglieri, R. Calderbank, A. Constantinides, A. Goldsmith, A. Paulraj and H. V. Poor, MIMO Wireless Communications, Cambridge University Press, Cambridge, 2007.